W9-CMI-643

An introduction to real and complex manifolds

Notes on mathematics and its applications

E. Artin ALGEBRAIC NUMBERS AND ALGEBRAIC FUNCTIONS
R. P. Boas COLLECTED WORKS OF HIDEHIKO YAMABE
R. A. Bonic LINEAR FUNCTIONAL ANALYSIS
M. Davis A FIRST COURSE IN FUNCTIONAL ANALYSIS
M. Davis LECTURES ON MODERN MATHEMATICS
J. Eells, Jr. SINGULARITIES OF SMOOTH MAPS
K. O. Friedrichs ADVANCED ORDINARY DIFFERENTIAL EQUATIONS
K. O. Friedrichs SPECIAL TOPICS IN FLUID DYNAMICS
K. O. Friedrichs and H. N. Shapiro INTEGRATION IN HILBERT SPACE
A. Guichardet SPECIAL TOPICS IN TOPOLOGICAL ALGEBRA
M. Hausner and J. T. Schwartz LIE GROUPS; LIE ALGEBRAS
P. Hilton HOMOTOPY THEORY AND DUALITY
F. John LECTURES ON ADVANCED NUMERICAL ANALYSIS
A. M. Krall STABILITY TECHNIQUES FOR CONTINUOUS LINEAR SYSTEMS
P. Lelong PLURISUBHARMONIC FUNCTIONS AND POSITIVE DIFFERENTIAL FORMS
H. Mullish AN INTRODUCTION TO COMPUTER PROGRAMMING
J. T. Schwartz DIFFERENTIAL GEOMETRY AND TOPOLOGY
J. T. Schwartz NONLINEAR FUNCTIONAL ANALYSIS
J. T. Schwartz *W-** ALGEBRAS
G. Sorani AN INTRODUCTION TO REAL AND COMPLEX MANIFOLDS
J. L. Soulé LINEAR OPERATORS IN HILBERT SPACE
J. J. Stoker NONLINEAR ELASTICITY

Additional volumes in preparation

An introduction to real and complex manifolds

Giuliano SORANI
Northeastern University
Boston, Mass.

GORDON AND BREACH
Science Publishers

NEW YORK LONDON PARIS

150 Fifth Avenue, New York, N.Y. 10011

Library of Congress catalog card number: 70–91367

Editorial office for the United Kingdom:
Gordon and Breach, Science Publishers Ltd.
12 Bloomsbury Way
London W.C.1

Editorial office for France:
Gordon & Breach
7–9 rue Emile Dubois
Paris 14e

Distributed in Canada by:
The Ryerson Press
299 Queen Street West
Toronto 2B, Ontario

to Viviana

Editors' preface

A LARGE NUMBER of mathematics books begin as lecture notes; but, because mathematicians are busy, and since the labor required to bring lecture notes up to the level of perfection which authors and the public demand of formally published books is very considerable, it follows that an even larger number of lecture notes make the transition to book form only after great delay or not at all. The present lecture note series aims to fill the resulting gap. It will consist of reprinted lecture notes, edited at least to a satisfactory level of completeness and intelligibility, though not necessarily to the perfection which is expected of a book. In addition to lecture notes, the series will include volumes of collected reprints of journal articles as current developments indicate, and mixed volumes including both notes and reprints.

JACOB T. SCHWARTZ
MAURICE LÉVY

Preface

THESE ARE the notes of a course I taught at Northeastern University during the winter and spring terms of 1967–68. The principal aim of the course was to provide the students with the greatest amount of information possible, while assuming the least amount of background. The topics covered were chosen to stimulate interest in the students taking the course rather than to involve them in the study of the subject at a sophisticated level. For this reason, the language of categories, which would have proven extremely useful in chapters six, seven, and eight, was avoided. The temptation to deal with deep theorems and to follow results to the best possible conclusions was also resisted.

Since the topics treated in these notes are classical, they have been borrowed from many sources. If someone is interested in investigating the origins of any of the material included, the bibliography at the end, although far from complete, may be helpful. In any case, the mistakes included represent my own contribution.

Some problems have been included at the end of every chapter to give the reader additional experience with the new techniques and practice in applying them. The reference accompanying each of the problems in parentheses denotes the section to which the problem is related. The results of all problems are freely used.

The system of enumeration employed should be easy to follow. For example, the notation (2,12) means number 12 of section 2 of the same chapter; the notation (3,2,12) means number 12 of section 2 of chapter 3. We shall use the sign ○ to indicate either the end of a proof or no proof.

I am indebted to all of the students who attended the course and provided healthy stimulation through their participation. Thanks are also due to those who carefully read the manuscript and suggested many improvements in it. Finally, I would like to express my gratitude to the National Science Foundation for its support during the summer of 1968, to Chris Robinson who typed the manuscript and to the staff of Gordon and Breach for their help in the preparation of this book.

GIULIANO SORANI

Contents

CHAPTER 1

Multilinear algebra

1.1 Vector spaces

WE SHALL DENOTE by $\mathbf{R}$ the field of real numbers and by $\mathbf{C}$ the field of complex numbers.

1.2 DEFINITION *A real vector space is a nonempty set V on which two operations:*

a) *addition*

$$+ : V \times V \to V$$

denoted by

$$(u, v) \to u + v \quad (u, v \in V),$$

and

b) *scalar multiplication*

$$\cdot : \mathbf{R} \times V \to V$$

denoted by

$$(c, u) \to cu \quad (c \in \mathbf{R},\ u \in V),$$

are defined and satisfy the following axioms:

i) *for all $u, v \in V$,* $\quad u + v = v + u$,

ii) *for all $u, v, w \in V$,* $\quad u + (v + w) = (u + v) + w$,

iii) *there exists an element $0 \in V$ such that*

$$0 + u = u$$

for all $u \in V$,

iv) *for every $u \in V$ there exists an element $-u \in V$ such that*

$$u + (-u) = 0,$$

v) *for all $u, v \in V$ and $c \in \mathbf{R}$*

$$c\,(u + v) = cu + cv,$$

vi) *for all $c, d \in \mathbf{R}$ and $u \in V$*

$$(c + d)\,u = cu + du,$$

vii) *for all $c, d \in \mathbf{R}$ and $u \in V$*

$$c\,(du) = (cd)\,u,$$

viii) *for all $u \in V$*

$$1u = u.$$

We remark that V is an abelian group under addition and that the element $0 \in V$ satisfying (iii) is unique. Also for every $u \in V$, the element $-u$ satisfying (iv) is unique.

A complex vector space is defined in the same way with **R** replaced by **C**.

We shall simply write vector space instead of real vector space. If V is a vector space, its elements will be called vectors. Elements of **R** (or **C**) will be called scalars.

1.3 Definition *A subset W of a vector space V is called a subspace of V if W is stable with respect to the operations* (a) *and* (b) *of V.*

Here, stable means that:

$$u, v \in W \quad \text{implies} \quad u + v \in W,$$

and

$$u \in W \quad \text{implies} \quad cu \in W \text{ for all } c \in \mathbf{R}.$$

Clearly W is itself a vector space.

A vector $v \in V$ is said to be a *finite linear combination* of the vectors $v_1, \ldots, v_k \in V$ if a relation

$$v = \sum_{i=1}^{k} c^i v_i$$

holds with $c^i \in \mathbf{R}$.

Let M be a subset of the vector space V. The set

$$L(M) = \left\{ \sum_{i=1}^{n} c^i v_i \mid v_i \in M, c^i \in \mathbf{R} \right\}$$

of all finite linear combinations of vectors of M is a subspace of V called the *subspace generated by M*, or the *span of M*. $L(M)$ is the smallest subspace of V containing M.

1.4 Definition *With the above notation, M is called a system of generators of V if $L(M) = V$.*

1.5 Definition *A vector space V is called finite dimensional if there exists in V a finite system $M = \{v_1, \ldots, v_k\}$ of generators.*

If n is the smallest number of generators of a finite dimensional vector space V, then we say that V has dimension n, and we write $\dim V = n$.

A vector space V is called *infinite dimensional* if V is not a finite dimensional vector space.

1.6 DEFINITION *A subset $F \subset V$ is said to be a free system of vectors if, for every finite subset $\{v_1, \ldots, v_k\} \subset F$, the relation:*

$$\sum_{i=1}^{k} c^i v_i = 0$$

implies $c^i = 0$ for $i = 1, \ldots, k$.

Vectors of a subset W of a vector space V are called *linearly independent* if W is a free system. They are called *linearly dependent* if W is not a free system; that is, if a relation of the type $\sum_i c^i v_i = 0$, $c^i \in \mathbf{R}$, $v_i \in W$, holds with some of the c^i's different from zero.

It is clear that every subset of a free system is itself a free system. In particular the empty set is free and every vector of a free system is free. Every vector $v \neq 0$ is free. The vector $0 \in V$ is linearly dependent.

If a subset $U \subset W$ is not free, then W too is not free. The vector $0 \in V$ cannot belong to a free system of vectors.

1.7 LEMMA *A subset $F \subset V$ is a free system of vectors if and only if no vector of F belongs to the subspace generated by the remaining vectors of F.* ○

1.8 LEMMA *Let $F \subset V$ be a free system of vectors and let $v \in V$ be a vector which does not belong to the subspace $L(F)$ generated by F. Then the set $F \cup \{v\}$ is a free system of vectors.* ○

The proof of the above lemmas is left as an exercise.

1.9 THEOREM *A vector space V has finite dimension n if and only if the maximum number of linearly independent vectors is $n < +\infty$.*

Proof We shall prove first that, if the maximum number of linearly independent vectors is $m < +\infty$, then V has dimension $n \leq m$. Let $F = \{v_1, \ldots, v_m\}$ be a free system of m vectors of V. By lemma 1.8, F is also a system of generators of V. Thus, by definition, $\dim V = n \leq m$.

We prove now that if $\dim V = n < +\infty$, then the maximum number of linearly independent vectors is $m \leq n$. Assume there is in V a free system $F = \{u_1, \ldots, u_m\}$ of $m > n$ vectors. We shall prove that $F' = \{u_1, \ldots, u_n\}$ is then a system of generators of V. Let $G = \{v_1, \ldots, v_n\}$ be any system of generators of V. Since u_1 is a linear combination of $v_1, \ldots, v_n$ it follows that for some j_1, $1 \leq j_1 \leq n$, v_{j_1} is a linear combination of $u_1, v_1, \ldots, \hat{v}_{j_1}, \ldots, v_n$. Therefore every vector $v \in V$ is a linear combination of the same vectors; that is, the set

$$G_1 = \{u_1, v_1, \ldots \hat{v}_{j_1}, \ldots, v_n\}$$

is also a system of generators of V. In the same way, since F is a free system, one proves that $G_2 = \{u_1, u_2, v_1, \ldots, \hat{v}_{j_1}, \ldots, \hat{v}_{j_2}, \ldots, v_n\}, \ldots, G_n = \{u_1, \ldots, u_n\} = F'$ are systems of generators. But then $F = \{u_1, \ldots, u_n, \ldots, u_m\}$ is not a free system contrary to the hypothesis. ○

From theorem 1.9 follows immediately:

1.10 Corollary *A vector space V is infinite dimensional if and only if V contains free systems $F = \{v_1, \ldots, v_n\}$ with n arbitrarily large.* ○

Let V be a vector space containing some vectors different from zero.

1.11 Definition *A basis of V is any subset $B \subset V$ which satisfies the following two conditions:*

i) *B is a free system of vectors of V,*

ii) *B is a system of generators of V.*

1.12 Theorem *Let V be a vector space of finite dimension n and let B be a subset of V. The following conditions are equivalent:*

a) *B is a basis of V,*

b) *B is a free system of n vectors,*

c) *B is a system of n generators of V.*

Proof By definition, (b) and (c) together imply (a). We shall show now that (a) implies (b) and (c). It suffices to prove that, if B is a basis of V, then B contains exactly n vectors. Since B is a free system, by theorem 1.9, B contains $m \leq n$ vectors. On the other hand, B is a system of generators, and therefore $m \geq n$. We prove now that (b) implies (c). Since $\dim V = n$, there exists a system of generators of V which contains exactly n vectors. By replacing the vectors in this system with those of B and using the same argument as in the proof of theorem 1.9 we show that B is a system of generators. We prove now that (c) implies (b). B consists of n generators. If B were not a free system, by lemma 1.7 there would exist a system of generators of V containing $m < n$ vectors. ○

Let V be a finite dimensional vector space, $\dim V = n$. Let $B = \{b_1, \ldots, b_n\}$ be a basis of V. If $v \in V$, then by (c) of theorem 1.12 we have:

$$v = \sum_{i=1}^{n} v^i b_i \quad (v^i \in \mathbf{R}).$$

By (b) of theorem 1.12 it follows that the coefficients v^i of the linear combination determine and are uniquely determined by v. They are called the *components* of v with respect to the basis B.

2.1 Examples

2.2 *Vectors of the three-dimensional affine space.* They are the equivalence classes of directed line segments with the same orientation and length. The set of these equivalence classes is a three-dimensional vector space. A basis of it is any set of three linearly independent vectors.

2.3 *The set* $\mathbf{R}_n[t]$ *of all polynomials in one indeterminate* t, *of degree* $\leq n-1$, *with real coefficients.* $\mathbf{R}_n[t]$ is a vector space of finite dimension n. A basis of $\mathbf{R}_n[t]$ is given by:

$$1, t, t^2, \ldots, t^{n-1}.$$

A different basis may be constructed as follows: Fix n arbitrary distinct points $t_0, \ldots, t_{n-1}$ and consider polynomials $p_k(t)$, $k = 0, 1, \ldots, n-1$, such that:

$$p_k(t) = \begin{cases} 1 & \text{if} \quad t = t_k \\ 0 & \text{if} \quad t = t_h \quad h \neq k. \end{cases}$$

Then $p_0(t), p_1(t), \ldots, p_{n-1}(t)$ form a basis of $\mathbf{R}_n[t]$.

2.4 *The set* $\mathbf{R}[t] = \bigcup_n \mathbf{R}_n[t]$. It is an infinite dimensional vector space with the basis:

$$1, t, t^2, \ldots .$$

A different basis of $\mathbf{R}[t]$ is constructed by considering an arbitrary sequence of points $\{t_k\}$ with $t_k \neq t_h$ if $k \neq h$, and by using the same polynomials as in example 2.3.

2.5 *The set* $\mathbf{R}[[t]]$ *of formal series in one indeterminate* t, *with real coefficients.* $\mathbf{R}[[t]]$ is an infinite dimensional vector space.

2.6 *The set* $\mathscr{C}^n(a, b)$, $(0 \leq n \leq +\infty)$ *of all functions of the real variable* x, *with continuous derivatives up to the order* n *in the interval* $(a, b) \subset \mathbf{R}$. With the usual operations:

$$(f+g)(x) = f(x) + g(x)$$

$$(cf)(x) = cf(x),$$

$\mathscr{C}^n(a, b)$ is an infinite dimensional vector space. In fact it contains $\mathbf{R}[t]$ as a subspace.

2.7 *The set* $L^2(a, b) = \left\{ f(x) \,\middle|\, \int_a^b |f(x)|^2 \, dx < +\infty \right\}$ *of all real (or complex) valued functions measurable and square integrable in* (a, b). With the same operations as in example 2.6, $L^2(a, b)$ is an infinite dimensional vector space.

3.1 Linear maps

Let V and W be two vector spaces. A map $T : V \to W$ is called a *linear map* (or a *vector space homomorphism*) if, for $v_1, v_2 \in V$,

$$T(v_1 + v_2) = T(v_1) + T(v_2),$$

and for $c \in \mathbf{R}$ and $v \in V$,

$$cT(v) = T(cv).$$

The two above conditions are clearly equivalent to:

$$T(c_1 v_1 + c_2 v_2) = c_1 T(v_1) + c_2 T(v_2)$$

$(c_1, c_2 \in \mathbf{R}; v_1, v_2 \in V)$.

If $S, T : V \to W$ are two linear maps we define $S = T$ if $S(v) = T(v)$ for every $v \in V$.

We shall denote by $\operatorname{Hom}(V, W)$ the set of all linear maps from V to W. The set $\operatorname{Hom}(V, W)$ is made into a vector space by the operations:

$$(S + T)(v) = S(v) + T(v)$$

$$(cT)(v) = cT(v),$$

$(S, T \in \operatorname{Hom}(V, W); v \in V; c \in \mathbf{R})$, which clearly satisfy the axioms of definition 1.2.

If V and W are finite dimensional vector spaces of dimensions n and m, then it is easy to verify that $\operatorname{Hom}(V, W)$ is of finite dimension nm. In particular, $\operatorname{Hom}(V, V)$ is n^2-dimensional if $\dim V = n$.

If V, W, Z are three vector spaces and $S \in \operatorname{Hom}(V, W)$, $T \in \operatorname{Hom}(W, Z)$, the composition $TS \in \operatorname{Hom}(V, Z)$ is defined by

$$TS(v) = T(S(v)), \quad v \in V.$$

If $T \in \operatorname{Hom}(V, W)$ the set:

$$\operatorname{Im} T = T(V) = \{w \in W \mid w = Tv \quad \text{for some} \quad v \in V\}$$

is a subspace of W, called the *image* of T.

A linear map T is called *surjective* if $\operatorname{Im} T = W$. The set

$$\operatorname{Ker} T = \{v \in V \mid Tv = 0 \in W\}$$

is a subspace of V, called the *kernel* of T. $\operatorname{Ker} T = T^{-1}(0)$.

A linear map T is called *injective* if $\operatorname{Ker} T = 0$.

Clearly T is an injective map if and only if T is one to one. If T is injective, there exists a linear map $T^{-1}\colon \operatorname{Im} T \to V$, the *inverse map*, such that $T^{-1}T =$ identity of V and $TT^{-1} =$ identity of $\operatorname{Im} T$ wherever these maps are defined.

3.2 The fact that a linear map T is surjective is equivalent to the fact that the equation in v:

3.3 $$Tv = w \quad (v \in V, w \in W)$$

always has a solution and every solution is of the form $v = v_1 + v_0$ where v_1 is a fixed solution of the equation 3.3 and v_0 is any element of $\operatorname{Ker} T$. T injective means that, if a solution of 3.3 exists, then it is unique.

A linear map T is called an *isomorphism* if T is injective and surjective; that is, if T is a bijection. If $T\colon V \to W$ is an isomorphism, then V and W are said to be *isomorphic*. If V is isomorphic to W we shall write $V \simeq W$.

The set of all isomorphisms of a vector space V onto itself is a group denoted by $GL(V)$.

3.4 *Remark* An isomorphism $T\colon V \to W$ induces an isomorphism: $\operatorname{Hom}(V, V) \to \operatorname{Hom}(W, W)$ defined by:

$$S \to TST^{-1} \quad (S \in \operatorname{Hom}(V, V)).$$

In particular, if $S \in GL(V)$, then

$$S \to TST^{-1}$$

is a group isomorphism: $GL(V) \to GL(W)$.

4.1 Dual spaces

Let V be a vector space. Since the set $\mathbf{R}$ is itself a vector space, we can consider $\operatorname{Hom}(V, \mathbf{R})$. This vector space plays an important role.

4.2 DEFINITION *The vector space* $\operatorname{Hom}(V, \mathbf{R})$ *is called the dual space of* V *and it is denoted by* V^*.

A linear map $v^* : V \to \mathbf{R}$ is called a *linear functional* on V. We shall denote by $\langle v, v^* \rangle$ or $v^*(v)$ the value of the linear functional $v^* \in V^*$ at the element $v \in V$. With this notation, the operations making Hom $(V, \mathbf{R})$ into a vector space are written as:

$$\langle v, v_1^* + v_2^* \rangle = \langle v, v_1^* \rangle + \langle v, v_2^* \rangle$$

and

$$\langle v, cv^* \rangle = c \langle v, v^* \rangle.$$

From the definition of the operations in Hom $(V, \mathbf{R})$ it appears that the map $v^* \to \langle v, v^* \rangle$ is a linear map $v^{**} \in$ Hom $(V^*, \mathbf{R})$. Therefore v^{**} is an element of the dual $(V^*)^* = V^{**}$ of V^*.

We thus have a map $J : V \to V^{**}$ with JV a subspace of V^{**}.

4.3 Definition *A vector space V is called algebraically reflexive if $JV = V^{**}$.*

We shall show later that all finite dimensional vector spaces are algebraically reflexive and vice-versa.

The pairing $\langle v, v^* \rangle$ is a "bilinear map" on $V \times V^*$ in the following sense: For fixed v^*, the map $v \to \langle v, v^* \rangle$ is a linear map $V \to \mathbf{R}$; for fixed v, the map $v^* \to \langle v, v^* \rangle$ is a linear map $V^* \to \mathbf{R}$.

Let V be a vector space of finite dimension n and let V^* be the dual of V. Let $B = \{b_i\}_{i=1,\ldots,n}$ be a basis of V. Let $b^{*i} \in V^*$ be the linear functional defined by:

$$\langle v, b^{*i} \rangle = v^i$$

if

$$v = \sum_{i=1}^{n} v^i b_i .$$

The linear functional b^{*i} is called the *coordinate functional* of index i with respect to the basis B of V. It follows from the definition that:

$$\langle b_i, b^{*j} \rangle = \delta_i^j \quad (i, j = 1, \ldots, n)$$

where δ_i^j is the Kronecker symbol:

$$\delta_i^j = \begin{cases} 1 & \text{if } i = j \\ 0 & \text{if } i \neq j. \end{cases}$$

4.4 Theorem *Let $B = \{b_1, \ldots, b_n\}$ be a basis of the vector space V. The n coordinate functionals $b^{*1}, \ldots, b^{*n}$ associated with the basis B form a basis of V^*.*

Proof We shall prove first that $B^* = \{b^{*1}, \ldots, b^{*n}\}$ is a system of generators of V^*. Let $v^* \in V^*$ and let $v = \sum_i v^i b_i$ be an element of V. Then $v^i = \langle v, b^{*i} \rangle$. We have:

$$\langle v, v^* \rangle = \langle \sum_i v^i b_i, v^* \rangle = \sum_i v^i \langle b_i, v^* \rangle$$

$$= \sum_i \langle v, b^{*i} \rangle \langle b_i, v^* \rangle = \langle v, \sum_i \langle b_i, v^* \rangle b^{*i} \rangle.$$

It follows that:

$$v^* = \sum_i v_i^* b^{*i}$$

with

$$v_i^* = \langle b_i, v^* \rangle.$$

We prove now that B^* is a free system of vectors. Let $\sum_i \alpha_i b^{*i} = 0$; then $\langle v, \sum_i \alpha_i b^{*i} \rangle = 0$ for every $v \in V$. By putting $v = b_j$, we get:

$$\langle b_j, \sum_i \alpha_i b^{*i} \rangle = \sum_i \alpha_i \langle b_j, b^{*i} \rangle = \alpha_j$$

which implies $\alpha_j = 0$ for all $j = 1, \ldots, n$. ○

4.5 Corollary *V^* has finite dimension n.* ○

4.6 Definition *With the same notation as in theorem* 4.4, *the basis B^* of V^* is called the dual basis of the basis B of V.*

4.7 *Remark* It is clear that the conditions:

$$\langle b_i, b^{*j} \rangle = \delta_i^j \quad (i, j = 1, \ldots, n)$$

characterize the basis B^* dual of B.

4.8 Theorem *A vector space V of finite dimension n is algebraically reflexive.*

Proof It suffices to show that, if B is a basis of V, then the image $B^{**} = JB$ of B in V^{**} is a basis of V^{**}. Now B^{**} is the dual basis of the basis B^* of V^* and

$$\langle b^{*i}, Jb_k \rangle = \langle b_k, b^{*i} \rangle = \delta_k^i. \quad ○$$

4.9 *Example* Dual space of the vector space $\mathbf{R}_n[t]$. The basis of $(\mathbf{R}_n[t])^*$, dual of the basis $1, t, \ldots, t^{n-1}$, is given by the linear functionals $f^j, j = 0, 1, \ldots, n-1$, defined by :

$$\langle p(t), f^j \rangle = p^{(j)}(0)/j!$$

where $p(t) \in \mathbf{R}_n[t]$ and $p^{(j)}(t)$ is the derived polynomial of order j of $p(t)$. Clearly $p^{(j)}(0) = j!\, a_j$ if $p(t) = \sum_i a_i t^i$.

The basis of $(\mathbf{R}_n[t])^*$ dual of the basis $\{p_k(t)\}_{k=0,\ldots,n-1}$ is given by the linear functionals f_k defined by:

$$\langle p(t), f_k \rangle = p(t_k) \quad (k = 0, 1, \ldots, n-1).$$

5.1 Infinite dimensional vector spaces

In order to show that an algebraically reflexive vector space is necessarily finite dimensional, we shall generalize theorem 1.12 to an arbitrary vector space and we shall show that every vector space V has a basis. If V is any vector space, the class of free systems of vectors and that of systems of generators of V are sets partially ordered by the inclusion relation.

5.2 THEOREM *Let V be any vector space and let B be a subset of V. The following properties are equivalent:*

i) *B is a basis of V,*
ii) *B is a maximal free system of V,*
iii) *B is a minimal system of generators of V.*

Proof We shall prove first that (i) implies (ii). Let B be a basis of V and let B' be a subset of V containing B as a proper subset. Then B' contains some elements which are not in B but belong to the subspace $L(B)$ generated by B. But since B is a basis, B' cannot be a free system by lemma 1.7.

We shall prove now that (i) implies (iii). Let B be a basis of V and let B'' be a proper subset of B. Let v be a vector of B, $v \notin B''$. Since B is a free system, by lemma 1.7, v does not belong to the subspace $L(B'')$ generated by B''. Therefore B'' is not a system of generators of V.

We prove now that (ii) implies (i). If B is a maximal free system, then the subspace $L(B)$ coincides with V. Otherwise there would exist, by lemma 1.8, a free system of vectors containing B as a proper subset.

We prove now that (iii) implies (i). If B is a minimal system of generators of V, then B is a free system. In fact, otherwise there would exist, by lemma 1.7, a vector $v \in B$ belonging to the subspace generated by the remaining vectors of B. Therefore B would contain, as a proper subset, a system of generators of V. ○

5.3 COROLLARY *Every vector space has a basis.*

Proof It is enough to show that the class of free systems of vectors of a vector space V has a maximal element. Since the union of all elements of a chain of free systems of V is a free system, the existence of a maximal free system of V follows by Zorn's lemma. ○

5.4 *Remark* If B is a basis of a subspace M of a vector space V, then there exists a basis B' of V which contains B. To prove this, it suffices to apply the same argument used in proving corollary 5.3 to the class of free systems of V which contain B.

5.5 THEOREM *A vector space V of infinite dimension is not algebraically reflexive.*

Proof We must show that JV is a proper subspace of V^{**}. Let $B = \{b_i\}_{i \in I}$ be a basis of V. For every $i \in I$ let $b^{*i} \in V^*$ be the coordinate functional defined by $\langle b_j, b^{*i}\rangle = \delta^i_j$. The set $\{b^{*i}\}_{i \in I}$ of coordinate functionals associated to the basis B is a free system of vectors of V^*. By remark 5.4 there exists a basis B' of V^* which contains this free system. Let K be the index set of B'. Every element $v^{**} \in V^{**}$ is determined by the values:

5.6
$$v_k^{**} = \langle b'^k, v^{**}\rangle, \; b'^k \in B',$$

taken by v^{**} at the vectors of B'. Conversely, given arbitrary v_k^{**}, $k \in K$, they determine a unique vector $v^{**} \in V^{**}$ which satisfies relations 5.6.

Now for every vector $v^{**} = Jv \in JV$ we have $v_i^{**} = \langle Jv, b^{*i}\rangle = v^i \neq 0$ only for a finite number of $i \in I$. On the other hand, there are evidently vectors $v^{**} \in V^{**}$ for which $v_i^{**} = \langle b^{*i}, v^{**}\rangle$ is different from zero for infinitely many values of $i \in I$. ○

5.7 *Remark* It is easy to show that a vector space V is algebraically reflexive if and only if the coordinate functionals associated to any basis of V form a basis of V^*. In fact, if V is algebraically reflexive, then V has finite dimension n, and V^* also has dimension n by theorem 4.4 and corollary 4.5. Therefore, by theorem 1.12, the coordinate functionals, being a free system of n vectors, are a basis of V^*.

Conversely, if the coordinate functionals $\{b^{*i}\}$, associated to the basis $\{b_i\}$ of V, are a basis of V^*, then the coordinate functionals $\{b_i^{**}\}$ of V^{**}, associated to the basis $\{b^{*i}\}$, are a basis of V^{**}. But then necessarily $V^{**} = V$, since the b_i^{**} are identified with the b_i:

$$\langle b^{*j}, b_i^{**}\rangle = \langle b_i, b^{*j}\rangle = \delta^j_i.$$

5.8 *Example* Let us consider again the vector space $\mathbf{R}[t]$ with the basis $1, t, t^2, \dots$. The coordinate functionals associated with this basis were denoted by f^j in example 4.9. Now it is clear that the f^j do not generate the whole of $(\mathbf{R}[t])^*$, as it is easy to check.

6.1 Properties of linear maps

The following theorems help to decide whether a linear map is injective or not.

6.2 THEOREM *Let $T \in \operatorname{Hom}(V, W)$. Then T is injective if and only if the image under T of every finite dimensional subspace of V is a finite dimensional subspace of TV of the same dimension.*

Proof If T takes every finite dimensional subspace of V into a subspace of the same dimension, then every finite dimensional subspace of $\operatorname{Ker} T$ has dimension zero and therefore $\operatorname{Ker} T = 0$.

Conversely, let M be a finite dimensional subspace of V. Since the image under T of a system of generators of M is a system of generators of TM, then $\dim TM \leq \dim M$. Now if $\operatorname{Ker} T = 0$, then $M = T^{-1}(TM)$ and $\dim M \leq \dim TM$. Thus, $\dim M = \dim TM$. ○

6.3 THEOREM *Let $T \in \operatorname{Hom}(V, W)$. Then T is injective if and only if the image under T of any free system of vectors of V is a free system of vectors of TV.*

Proof Let F be a free system of V. Since T is injective, $F = T^{-1}(TF)$. Then TF must be a free system of TV; otherwise F would not be a free system.

Conversely, if T takes free systems into free systems and $v \neq 0$ is a vector of V, we cannot have $Tv = 0$ since v is a free system while 0 is not a free system. ○

6.4 THEOREM *Let $T \in \operatorname{Hom}(V, W)$. Then T is injective if and only if the image under T of any basis of V is a basis of TV.*

Proof Let B be a basis of V and let T be injective. Then B is a maximal free system of V. Therefore TB is a free system by theorem 6.3 and it is maximal by theorem 6.2. Thus TB is a basis of TV by theorem 5.2.

Conversely, let F be a free system of vectors and let $F \subset B$ where B is a basis of V. If T takes bases of V into bases of TV, then TB is a basis of TV.

Therefore TB is a free system of vectors and $TF \subset TB$ is also a free system of vectors. Then T is injective by theorem 6.3. ○

6.5 THEOREM *Let* $T \in \text{Hom}(V, W)$, *and let* $\dim V = \dim W = n < +\infty$. *Then T is injective if and only if T is surjective.*

Proof If T is injective, it follows from theorem 6.2 that $\dim TV = \dim V = \dim W$ and then $TV = W$.

Conversely, let B be a basis of V. Then TB is a system of generators of TV and contains at most n elements. But since $TV = W$, then TB must contain exactly n elements and therefore TB is a basis of W, by theorem 1.12. Thus T takes bases of V onto bases of W and therefore T is injective by theorem 6.4. ○

6.6 COROLLARY *Two finite dimensional vector spaces are isomorphic if and only if they have the same dimension.*

Proof Let V and W be two isomorphic finite dimensional vector spaces. There exists an injective map of V onto W and then by theorem 6.2 $\dim V = \dim W$.

Conversely, let $\dim V = \dim W = n < +\infty$ and let $\{v_1, \ldots, v_n\}$ and $\{w_1, \ldots, w_n\}$ be two bases of V and W. The linear map:

$$\alpha : \sum_i c^i v_i \to \sum_i c^i w_i \quad (c^i \in \mathbf{R})$$

is evidently an isomorphism of V onto W. ○

6.7 *Remark* The isomorphism $\alpha : V \to W$ depends on the choice of the bases of V and W.

Let V, W be two vector spaces and let V^*, W^* be their dual spaces. If $T \in \text{Hom}(V, W)$, there is an element $T^* \in \text{Hom}(W^*, V^*)$ uniquely defined by:

$$\langle Tv, w^* \rangle = \langle v, T^* w^* \rangle,$$

$v \in V$, $w^* \in W^*$.

In fact for every fixed $w^* \in W^*$, the map $v \to \langle Tv, w^* \rangle$ is a linear functional v^* on V. We put $v^* = T^* w^*$. Obviously $T^* \in \text{Hom}(W^*, V^*)$. The linear map T^* is called the *transpose* of T.

The map $T \to T^*$ is a linear map: $\text{Hom}(V, W) \to \text{Hom}(W^*, V^*)$, in the sense that

$$(c_1 T_1 + c_2 T_2)^* = c_1 T_1^* + c_2 T_2^*,$$

$c_1, c_2 \in \mathbf{R}$; $T_1, T_2 \in \text{Hom}(V, W)$.

The transpose map satisfies the following properties which are easy to check:

i) Let $S \in \text{Hom}(V, W)$ and $T \in \text{Hom}(W, Z)$, then $S^* \in \text{Hom}(W^*, V^*)$ and $T^* \in \text{Hom}(Z^*, W^*)$. We have:

$$(TS)^* = S^*T^*.$$

ii) If $T \in \text{Hom}(V, W)$ is an isomorphism, so is $T^* \in \text{Hom}(W^*, V^*)$ and $(T^{-1})^* = (T^*)^{-1}$.

iii) If V and W are finite dimensional vector spaces and $T \in \text{Hom}(V, W)$, then $(T^*)^* = T^{**} = T$.

7.1 More examples

In the following, we shall make constant use of some particular vector spaces which we describe here.

7.2 The set $\mathbf{R}^n$ of all ordered n-tuples of real numbers is a vector space of finite dimension n with operations:

$$x + y = \{x_1, \ldots, x_n\} + \{y_1, \ldots, y_n\} = \{x_1 + y_1, \ldots, x_n + y_n\}$$

$$cx = c\,\{x_1, \ldots, x_n\} = \{cx_1, \ldots, cx_n\}$$

$(x, y \in \mathbf{R}^n;\ c \in \mathbf{R})$.

This vector space is denoted by V^n. A vector $v = \{x_i\} \in V^n$ determines the n-tuple $\{v^i\}$ of its components with respect to every choice of a basis of V^n. It is easy to check that the n vectors:

7.3
$$\begin{cases} e_1 = \{1, 0, \ldots, 0\} \\ e_2 = \{0, 1, 0, \ldots, 0\} \\ \cdots\cdots\cdots\cdots \\ e_n = \{0, \ldots, 0, 1\} \end{cases}$$

form a basis B_0 of V^n such that for every $v \in V^n$, $v^i = x_i$. B_0 is called the *canonical* basis of V^n.

7.4 Let X be any set and let $H(X, \mathbf{R})$ be the set of all maps $f\colon X \to \mathbf{R}$. If $f, g \in H(X, \mathbf{R})$ then $f = g$ if $f(x) = g(x)$ for all $x \in X$. For $f, g \in H(X, \mathbf{R})$ and $c \in \mathbf{R}$ define:

$$(f + g)(x) = f(x) + g(x)$$

and
$$(cf)(x) = cf(x).$$

Then $f + g$ and cf are elements of $H(X, \mathbf{R})$ and these definitions make $H(X, \mathbf{R})$ into a vector space.

If we denote by $H_0(X, \mathbf{R})$ the set of all $f \in H(X, \mathbf{R})$ such that $f(x) = 0$ except for a finite number of $x \in X$, it is easy to show that $H_0(X, \mathbf{R})$ is also a vector space.

Vector spaces of this example may be of finite or infinite dimension.

7.5 The set $M(m, n)$ of all $m \times n$ matrices with real entries is a vector space of finite dimension mn with the operations:

$$A + B = (a^i_j + b^i_j),$$
$$cA = (ca^i_j),$$

$(A = (a^i_j),\ B = (b^i_j) \in M(m, n);\ c \in \mathbf{R};\ i = 1, \ldots, m;\ j = 1, \ldots, n).$

If $A \in M(m, n)$ and $\{B^i_j\}^{i=1,\ldots,m}_{j=1,\ldots,n}$ is a basis of $M(m, n)$, we shall denote by A^i_j the components of A with respect to that basis. It is easily seen that the matrices $\Delta^h_k = (\delta^h_i \delta^j_k)$ form a basis of $M(m, n)$ such that for every $A = (a^i_j) \in M(m, n)$, $A^i_j = a^i_j$.

7.6 *Remark* The set of all $n \times n$ non-degenerate matrices, that is, with determinant different from zero, with the usual product of matrices, form a group which is denoted by $GL(n, \mathbf{R})$, the general linear group.

8.1 Transformations

If V is any vector space of finite dimension n, then, by corollary 6.6, V is isomorphic to the vector space V^n. Moreover, whenever we fix a basis B of V, then to every vector $v \in V$ is associated the n-tuple $\{v^i\} \in \mathbf{R}^n$ of its components with respect to the basis B.

In other words, the choice of a basis B of V assigns an isomorphism:

$$\alpha_B : V \to V^n$$

which takes the basis B onto the basis B_0 of V^n.

Conversely, if $T: V \to V^n$ is an isomorphism, then there exists a basis B of V such that to every $v \in V$ the isomorphism T associates the n-tuple of its components with respect to the basis B. Clearly that basis is given by $T^{-1}(B_0)$.

Now let V be an n-dimensional vector space and W be an m-dimensional vector space. Let $B = \{b_1, \ldots, b_n\}$ and $C = \{c_1, \ldots, c_m\}$ be bases of V and W. To every $v \in V$ and $w \in W$ we associate the n-tuple $\{v^i\}$ and the m-tuple $\{w^j\}$ of their components with respect to the given bases B and C. This induces an isomorphism of the vector space Hom (V, W) onto the vector space $M(m, n)$. To every element $T \in \text{Hom}(V, W)$ corresponds the matrix (a_i^j) such that $w = Tv$ is equivalent to

$$w^j = \sum_{i=1}^{n} a_i^j v^i; \quad j = 1, \ldots, m.$$

In particular, let us consider an element $T \in \text{Hom}(V, V)$. Let $\{b_i\}_{i=1,\ldots,n}$ be a basis of the first copy of V and let $\{c_i\}_{i=1,\ldots,n}$ be a basis of the second copy of V. Let $w = Tv$. Then it is easy to check that for the corresponding matrix $(a_i^j) \in M(m, n)$ we have:

8.2 $$a_i^j = \langle Tb_i, c^{*j} \rangle,$$

where $\{c^{*j}\}$ is the dual basis of the basis $\{c_i\}$. In fact, one has:

$$w = \sum_{i=1}^{n} w^i c_i = T \sum_{i=1}^{n} v^i b_i = \sum_{i=1}^{n} v^i T b_i.$$

Then 8.2 follows, by applying the functional c^{*j}.

8.3 *Remark* If $T \in \text{Hom}(V, V)$ is an isomorphism, then the matrix (a_i^j), associated to T in a fixed basis, is nondegenerate. Thus the isomorphism:

$$\text{Hom}(V, V) \to M(n, n)$$

induces an isomorphism:

$$GL(V) \to GL(n, \mathbf{R}).$$

8.4 Now let $T \in \text{Hom}(V, W)$ and let T^* be the transpose map of T. Let $v^* \in V^*$ and $w^* \in W^*$. With respect to the bases $\{b^{*i}\}$, $\{c^{*j}\}$ dual of the bases $\{b_i\}$, $\{c_j\}$ one has $v^* = \sum_{i=1}^{n} v_i^* b^{*i}$ and $w^* = \sum_{j=1}^{m} w_j^* c^{*j}$. With respect to the same bases the linear map $T^* \in \text{Hom}(W^*, V^*)$ is represented by the matrix $(\tilde{a}_i^j)$ such that, if $v^* = T^* w^*$, then $v_i^* = \sum_{j=1}^{m} \tilde{a}_i^j w_j^*$, $i = 1, \ldots, n$.

If $A = (a_j^i)$ is the matrix associated to T, then one has $\tilde{a}_i^j = a_j^i$, $i = 1, \ldots, n$; $j = 1, \ldots, m$. The matrix $(\tilde{a}_i^j)$ is called the *transpose* of the matrix A and denoted by ${}^tA$. Thus $A \in M(m, n)$, ${}^tA \in M(n, m)$.

In particular, if $T \in \operatorname{Hom}(V, V)$, both (a^i_j) and ${}^t(a^i_j)$ are elements of $M(n, n)$. Only the role of rows and columns is interchanged. With the same notation as in 8.1, one has:

$$a^i_j = \tilde{a}^j_i = \langle Tb_j, c^{*i} \rangle = \langle b_j, T^* c^{*i} \rangle.$$

8.5 Let V be an n-dimensional vector space; let $B_1 = \{b_{1i}\}$, $B_2 = \{b_{2i}\}$ be two bases of V.

Let us consider the matrix $B = (\langle b_{1i}, b_2^{*j} \rangle)$. Since

$$\langle b_{1i}, b_2^{*j} \rangle = \sum_{h=1}^{n} \langle b_{1h}, b_2^{*j} \rangle \, \delta^h_i,$$

B defines an isomorphism $\alpha_{B_1 B_2} : V \to V$ which takes the basis B_2 onto the basis B_1. On the other hand, if $v \in V$ and $v = \sum_{i=1}^{n} v^i_1 b_{1i} = \sum_{i=1}^{n} v^i_2 b_{2i}$, then by applying b_2^{*j}, one has $\langle v, b_2^{*j} \rangle = \sum_{i=1}^{n} v^i_1 \langle b_{1i}, b_2^{*j} \rangle$ and, since $\langle v, b_2^{*j} \rangle = v^j_2$, it follows that the matrix B transforms the components of a vector v with respect to the basis B_1 into the components of the same vector v with respect to the basis B_2.

The matrix B is obviously nondegenerate and therefore it is an element of $GL(n, \mathbf{R})$. Conversely, whenever we fix a basis $\{b_i\}$ of V, every element $A \in GL(n, \mathbf{R})$ determines the basis $\{c_i\}$ of V such that $a^j_i = \langle b_i, c^{*j} \rangle$.

Let $T \in \operatorname{Hom}(V, V)$ be an isomorphism and let (a^i_{1j}), (a^i_{2j}) be the matrices associated to T in the bases B_1 and B_2. Now since $\alpha_{B_2} \circ \alpha_{B_1}^{-1} = \alpha_{B_2 B_1}$, from the diagram:

$$\begin{array}{ccccc} V^n & \xrightarrow{\alpha_{B_1}^{-1}} & V & \xrightarrow{\alpha_{B_2}} & V^n \\ \downarrow (a^i_{1j}) & & \downarrow T & & \downarrow (a^i_{2j}) \\ V^n & \xrightarrow{\alpha_{B_1}^{-1}} & V & \xrightarrow{\alpha_{B_2}} & V^n \end{array}$$

it follows that:

$$(a^i_{2j}) = \alpha_{B_2 B_1} \circ (a^i_{1j}) \circ (\alpha_{B_2 B_1})^{-1},$$

and, since $(\alpha_{B_2 B_1})^{-1} = \alpha_{B_1 B_2}$, one has:

$$a^i_{2j} = \sum_{h,k=1}^{n} \langle b_{2j}, b_1^{*k} \rangle \, a^h_{1k} \, \langle b_{1h}, b_2^{*i} \rangle.$$

9.1 Direct sum and quotient space

Let V, W be two vector spaces. The direct sum $V \oplus W$ of V and W is the vector space consisting of all ordered pairs (v, w), $v \in V$, $w \in W$, with operations defined by:

$$(v, w) + (v', w') = (v + v', w + w')$$

$$c\,(v, w) = (cv, cw),$$

for $v, v' \in V$; $w, w' \in W$; $c \in \mathbf{R}$.

The direct sum $V \oplus W$ of two vector spaces V, W gives rise to a diagram:

$$V \underset{p}{\overset{i}{\rightleftarrows}} V \oplus W \underset{q}{\overset{j}{\leftrightarrows}} W$$

where the linear maps i, j, p, q are defined by $i(v) = (v, 0)$, $j(w) = (0, w)$, $p\,(v, w) = v$, $q\,(v, w) = w$. The maps i, j are called the *injections* and the maps p, q are called the *projections* of the direct sum.

The direct sum we defined is usually called the *external direct sum.*

Let V be a vector space and let W be a subspace of V. For $v \in V$, the set:

$$v + W = \{v + w \mid w \in W\}$$

is called a *coset* of W. Two such cosets are either disjoint or identical. Let V/W be the set of all cosets of W. Now for $v + W, v' + W \in V/W$ and $c \in \mathbf{R}$ define

$$(v + W) + (v' + W) = (v + v') + W$$

and

$$c\,(v + W) = cv + W.$$

We leave it to the reader to verify that these operations are well defined and that they actually make V/W into a vector space. This vector space V/W is called the *quotient space* of V by W. The map $V \to V/W$ sending every $v \in V$ into the coset $v + W$ is called the *natural projection.*

10.1 Tensor product

Let V, W, M be vector spaces. A map:

$$T : V \times W \to M$$

is called *bilinear* if, for fixed $v \in V, T$ is a linear map $W \to M$, and, for fixed $w \in W$, T is a linear map $V \to M$.

Let (M, h) be a pair consisting of a vector space M and a bilinear map $h: V \times W \to M$ such that:

i) $h(V \times W)$ generates M,

ii) to any vector space N and bilinear map $f: V \times W \to N$ there is a unique linear map $\phi: M \to N$ such that the diagram:

$$\begin{array}{ccc} V \times W & & \\ \downarrow h & \searrow f & \\ M & \xrightarrow{\phi} & N \end{array}$$

commutes.

10.2 DEFINITION *Any pair* (M, h) *satisfying conditions* (i) *and* (ii) *is called a tensor product of* V *and* W.

That a tensor product is unique up to an isomorphism is shown by the next theorem.

10.3 THEOREM *If* (M, h) *and* (M', h') *are two tensor products of the vector spaces* V *and* W, *there is an isomorphism* α *of* M *onto* M' *such that* $\alpha \circ h = h'$.

Proof Property (ii) shows the existence of linear maps $\alpha: M \to M'$ and $\alpha': M' \to M$ with $\alpha \circ h = h'$ and $\alpha' \circ h' = h$. It follows that $\alpha \circ \alpha' =$ identity on $h'(V \times W)$ and thus on M' by (i). Similarly $\alpha' \circ \alpha =$ identity on M. ○

We shall now construct an example of a tensor product of two vector spaces V, W. Let $V \cdot W$ be the set of all formal sums

$$\Sigma\, a_{vw}\,(v, w)$$

where $v \in V$, $w \in W$, $a_{vw} \in \mathbf{R}$ and $a_{vw} \neq 0$ only for a finite number of pairs $(v, w) \in V \times W$. It is easy to see that $V \cdot W$ is a vector space.

Let $\mathscr{R}$ be the subspace of $V \cdot W$ generated by all elements of the form:

$$(a_1 v_1 + a_2 v_2, b_1 w_1 + b_2 w_2) - a_1 b_1 (v_1, w_1) - a_1 b_2 (v_1, w_2)$$
$$- a_2 b_1 (v_2, w_1) - a_2 b_2 (v_2, w_2).$$

If we identify (v, w) with $1\,(v, w)$, then $V \times W$ becomes a subset of $V \cdot W$.

Define $V \otimes W = V \cdot W / \mathscr{R}$. Let $h': V \cdot W \to V \otimes W$ be the natural projection and let $h: V \times W \to V \otimes W$ be the restriction of h' to the subset $V \times W$. We shall prove that the pair $(V \otimes W, h)$ is a tensor product of V and W.

We prove first that h is bilinear on $V \times W$. The formal sums:

$$(a_1 v_1 + a_2 v_2, w) - a_1 (v_1, w) - a_2 (v_2, w)$$

$$(v, b_1 w_1 + b_2 w_2) - b_1 (v, w_1) - b_2 (v, w_2)$$

belong to $\mathscr{R}$. Therefore their image under h is zero. Thus:

$$h (a_1 v_1 + a_2 v_2, w) = a_1 h (v_1, w) + a_2 h (v_2, w)$$

$$h (v, b_1 w_1 + b_2 w_2) = b_1 h (v, w_1) + b_2 h (v, w_2),$$

which proves that h is bilinear.

Since property (i) follows from the definition, we have only to prove that (ii) is valid. Let $f\colon V \times W \to N$ be any bilinear map. Define $\hat{f}\colon V \cdot W \to N$ by $\hat{f} \Sigma a_{vw}(v, w) = \Sigma a_{vw} f(v, w)$. It is easy to check that $\hat{f}$ maps $\mathscr{R}$ onto zero and then induces a linear map $\tilde{f}\colon V \otimes W \to N$ such that:

$$\tilde{f} h (v, w) = \tilde{f}((v, w) + \mathscr{R}) = \hat{f}(v, w) = f(v, w)$$

for all elements (v, w) of $V \times W$.

The pair $(V \otimes W, h)$ which we have just constructed is called *the tensor product* of V and W and will simply be denoted by $V \otimes W$. If $W = V$, we shall also write $\otimes^2 V$ for $V \otimes V$. If $v \in V$, $w \in W$, the element $h(v, w)$ of $V \otimes W$ will be denoted by $v \otimes w$.

If V, W, M are vector spaces, we shall denote by $\mathrm{Hom}(V, W; M)$ the vector space of all bilinear maps $V \times W \to M$.

10.4 THEOREM $\mathrm{Hom}(V, W; M)$ *is isomorphic to* $\mathrm{Hom}(V \otimes W, M)$.

Proof Given a linear map $g\colon V \otimes W \to M$, the map $f = g \circ h\colon V \times W \to M$ is bilinear. Thus we can define a linear map:

$$\mathrm{Hom}(V \otimes W, M) \to \mathrm{Hom}(V, W; M)$$

which is surjective by definition of $V \otimes W$. Moreover, if $g \circ h = 0$, it follows that $g = 0$ since $h(V \times W)$ generates $V \otimes W$. ○

By taking $M = \mathbf{R}$ in the above theorem, it follows that $(V \otimes W)^*$ is isomorphic to the space of bilinear maps of $V \times W$ into $\mathbf{R}$.

10.5 THEOREM *Let V and W be finite dimensional vector spaces. If* $\dim V = n$ *and* $\dim W = m$, *then* $\dim V \otimes W = nm$. *If $B = \{b_i\}_{i=1,\dots,n}$ and $G = \{g_j\}_{j=1,\dots,m}$ are bases of V and W respectively, then $\{b_i \otimes g_j\}$ is a basis of $V \otimes W$.*

Proof Define $\phi_{lk}: V \times W \to \mathbf{R}$ by

$$\phi_{lk}(b_i, g_j) = \delta_i^l \delta_j^k \qquad l, i = 1, \ldots, n;$$
$$k, j = 1, \ldots, m.$$

The maps ϕ_{lk} extend by linearity to any element of $V \times W$ and are linearly independent in Hom $(V, W; \mathbf{R})$. In fact, if $\sum_{l,k} a_{lk}\phi_{lk} = 0$, $a_{lk} \in \mathbf{R}$, then we have

$$0 = \sum_{l,k} a_{lk}\phi_{lk}(b_i, g_j) = \sum_{l,k} a_{lk}\delta_i^l\delta_j^k = a_{ij}.$$

Therefore, $\dim (V \otimes W)^* = \dim \operatorname{Hom}(V, W; \mathbf{R}) \geq nm$. Now every element $v \otimes w = h(v, w)$ is a linear combination of $b_i \otimes g_j = h(b_i, g_j)$, and since $h(V \times W)$ generates $V \otimes W$, it follows that the $b_i \otimes g_j$ generate $V \otimes W$. Thus $\dim (V \otimes W)^* = \dim V \otimes W \leq nm$. ○

11.1 Tensors

Let V be a vector space of finite dimension n. A p-linear form Φ on V is a real valued function $\Phi(v_1, \ldots, v_p)$ of $p \geq 1$ arguments $v_1, \ldots, v_p \in V$, which is linear with respect to each of them:

$$\Phi(v_1, \ldots, av_i + bw_i, \ldots, v_p) = a\Phi(v_1, \ldots, v_i, \ldots, v_p)$$
$$+ b\Phi(v_1, \ldots, w_i, \ldots, v_p),$$

$(v_i, w_i \in V; a, b \in \mathbf{R})$.

A p-linear form on V is called *a covariant tensor of order p on V*. In particular, a covariant tensor of order one is just a linear functional on V.

The set of all covariant tensors of order p on V, with the operations:

$$(\Phi_1 + \Phi_2)(v) = \Phi_1(v) + \Phi_2(v)$$
$$(c\Phi)(v) = c\Phi(v),$$

where $v = (v_1, \ldots, v_p)$ and $c \in \mathbf{R}$, is called the p-th tensor power of V^*. In fact, it follows from theorem 10.4, which can be easily generalized (see problem 14), that this space is isomorphic to $\otimes^p V^*$. Obviously one has $\otimes^1 V^* = V^*$.

Since V can be identified with the dual of V^*, a covariant tensor of order p on V^* will be called a *contravariant tensor of order p on V*.

The vector space $\otimes^p V^{**}$ of contravariant tensors of order p on V is denoted by $\otimes^p V$.

For $p, q \geq 1$ a multiplication

$$\otimes^p V^* \times \otimes^q V^* \to \otimes^{p+q} V^*$$

is defined by

$$\Phi \otimes \psi (v_1, \ldots, v_p, w_1, \ldots, w_q) = \Phi (v_1, \ldots, v_p) \psi (w_1, \ldots, w_q),$$

$v_i, w_j \in V$.

$\Phi \otimes \psi \in \otimes^{p+q} V^*$ is called the *tensor product* of Φ and ψ. This product satisfies the distributive laws in the sense that, for $\Phi, \Phi' \in \otimes^p V^*$ and $\psi, \psi' \in \otimes^q V^*$, one has:

$$(\Phi + \Phi') \otimes \psi = \Phi \otimes \psi + \Phi' \otimes \psi,$$

$$\Phi \otimes (\psi + \psi') = \Phi \otimes \psi + \Phi \otimes \psi'.$$

Moreover, as it is easy to check, for $\Phi \in \otimes^p V^*$, $\psi \in \otimes^q V^*$, $c \in \mathbf{R}$, one has:

$$(c\Phi) \otimes \psi = \Phi \otimes (c\psi) = c\Phi \otimes \psi.$$

Let V be a vector space, let $\{b_i\}$ be a basis of V, and let $\{b^{*i}\}$ be the dual basis of V^*. The tensor product $b^{*i} \otimes b^{*j}$ is an element of $\otimes^2 V^*$. By definition, for $v_1, v_2 \in V$, one has:

$$b^{*i} \otimes b^{*j} (v_1, v_2) = \langle v_1, b^{*i} \rangle \langle v_2, b^{*j} \rangle.$$

More generally, if $v_1, \ldots, v_p \in V$ and $b^{*i_1} \otimes \cdots \otimes b^{*i_p} \in \otimes^p V^*$, one has: $b^{*i_1} \otimes \cdots \otimes b^{*i_p} (v_1, \ldots, v_p) = \langle v_1, b^{*i_1} \rangle \cdots \langle v_p, b^{*i_p} \rangle$.

For simplicity of notation we shall prove the next lemma for $p = 2$.

11.2 LEMMA *Let $\{b_i\}_{i=1,\ldots,n}$ be a basis of the vector space V, $\{b^{*j}\}_{j=1,\ldots,n}$ its dual basis of V^*. Then $\otimes^2 V^*$ is a vector space of dimension n^2. A basis of $\otimes^2 V^*$ is given by the n^2 tensor products $b^{*i} \otimes b^{*j}$, $(i, j = 1, \ldots, n)$.*

Proof Let $v_1 = \sum_{i=1}^n v_1^i b_i$, $v_2 = \sum_{j=1}^n v_2^j b_j$ be vectors of V and let $\Phi \in \otimes^2 V^*$. Then:

$$\Phi (v_1, v_2) = \Phi \left(\sum_{i=1}^n v_1^i b_i, \sum_{j=1}^n v_2^j b_j \right) = \sum_{i,j=1}^n v_1^i v_2^j \Phi (b_i, b_j)$$

$$= \sum_{i,j=1}^n \Phi (b_i, b_j) \langle v_1, b^{*i} \rangle \langle v_2, b^{*j} \rangle$$

$$= \sum_{i,j=1}^n \Phi (b_i, b_j) b^{*i} \otimes b^{*j} (v_1, v_2).$$

It follows that the elements $b^{*i} \otimes b^{*j}$, $(i, j = 1, \dots, n)$, generate $\otimes^2 V^*$. Moreover, if

$$\sum_{i,j=1}^{n} c_{ij} b^{*i} \otimes b^{*j} = 0,$$

then we have:

$$0 = \sum_{i,j=1}^{n} c_{ij} b^{*i} \otimes b^{*j} (b_h, b_k) = \sum_{i,j=1}^{n} c_{ij} \langle b_h, b^{*i} \rangle \langle b_k, b^{*j} \rangle$$

$$= \sum_{i,j=1}^{n} c_{ij} \delta^i_h \delta^j_k = c_{hk},$$

and therefore the n^2 tensor products $b^{*i} \otimes b^{*j}$ form a basis of $\otimes^2 V^*$. ○

In the same way it can be proved that the n^p tensor products of the b^{*i}, p by p form a basis of $\otimes^p V^*$. It follows that, for every $\Phi \in \otimes^p V^*$, one has:

$$\Phi = \sum_{i_1 \cdots i_p} \Phi_{i_1 \cdots i_p} b^{*i_1} \otimes \cdots \otimes b^{*i_p}, (1 \leq i_1 \cdots i_p \leq n)$$

where $\Phi_{i_1 \cdots i_p} = \Phi(b_{i_1}, \dots, b_{i_p})$ are the components of the tensor Φ with respect to the basis $\{b^{*i_1} \otimes \cdots \otimes b^{*i_p}\}$.

Now let $\{c_i\}_{i=1,\dots,n}$ be a second basis of V and let $(a^j_i) = (\langle c_i, b^{*j} \rangle)$ be the matrix which takes the basis $\{b_i\}$ into the basis $\{c_i\}$. With respect to the two bases induced by $\{b_i\}$ and $\{c_i\}$ in $\otimes^p V^*$, the components of a tensor $\Phi \in \otimes^p V^*$ are related by the formula:

$$\Phi_{2\, j_1 \cdots j_p} = \sum_{i_1 \cdots i_p} a^{i_1}_{j_1} \cdots a^{i_p}_{j_p} \Phi_{1\, i_1 \cdots i_p},$$

where we denote by Φ_1, Φ_2 the components of Φ with respect to the bases $\{b_i\}$, $\{c_i\}$.

11.3 *Remark* Since a contravariant tensor on V is just a covariant tensor on V^*, it is clear that the above theory applies to contravariant tensors as well. For example, a basis $\{b_i\}$ of V induces the basis $b_{i_1} \otimes \cdots \otimes b_{i_p}$, $(1 \leq i_1 \cdots i_p \leq n)$ of $\otimes^p V$ and, if $\Phi \in \otimes^p V$, then we have

$$\Phi = \sum_{i_1 \cdots i_p} \Phi^{i_1 \cdots i_p} b_{i_1} \otimes \cdots \otimes b_{i_p}$$

where $\Phi^{i_1 \cdots i_p} = \Phi(b^{*i_1}, \dots, b^{*i_p})$ are the components of Φ with respect to the basis $\{b_{i_1} \otimes \cdots \otimes b_{i_p}\}$ of $\otimes^p V$. And, if $\{c_i\}$ is a second basis of V, with the same notation as before, we have

$$\Phi^{2\, j_1 \cdots j_p} = \sum_{i_1 \cdots i_p} a^{j_1}_{i_1} \cdots a^{j_p}_{i_p} \Phi^{1\, i_1 \cdots i_p}$$

where $a^j_i = \langle b_i, c^{*j} \rangle$.

12.1 Symmetric and antisymmetric tensors

Assume first $p = 2$ and consider the space $\otimes^2 V^*$ of all covariant tensors of order 2 on V. Define linear maps:

$$A : \otimes^2 V^* \to \otimes^2 V^*$$

$$S : \otimes^2 V^* \to \otimes^2 V^*$$

by

$$A\Phi\,(v_1, v_2) = \tfrac{1}{2}\,[\Phi\,(v_1, v_2) - \Phi\,(v_2, v_1)],$$

$$S\Phi\,(v_1, v_2) = \tfrac{1}{2}\,[\Phi\,(v_1, v_2) + \Phi\,(v_2, v_1)],$$

for $\Phi \in \otimes^2 V^*$; $v_1, v_2 \in V$.

It is easy to verify that A and S satisfy the following properties:

$$A^2 = A, \quad S^2 = S, \quad A + S = \text{identity}, \quad AS = SA = 0.$$

It follows that the space $\otimes^2 V^*$ splits into a direct sum:

$$\otimes^2 V^* = A\otimes^2 V^* \oplus S\otimes^2 V^*.$$

The elements of the subspace $A\otimes^2 V^*$, which are characterized by the condition $A\Phi = \Phi$, are called *antisymmetric tensors*. The elements of the subspace $S\otimes^2 V^*$, which are characterized by the condition $S\Phi = \Phi$, are called *symmetric tensors*.

Now, for any p, the linear maps $A : \otimes^p V^* \to \otimes^p V^*$ and $S : \otimes^p V^* \to \otimes^p V^*$ are defined by:

$$A\Phi\,(v_1, \ldots, v_p) = 1/p! \sum_{i_1 \cdots i_p} \operatorname{sign}\,(i_1, \ldots, i_p)\, \Phi\,(v_{i_1}, \ldots, v_{i_p})$$

$$S\Phi\,(v_1, \ldots, v_p) = 1/p! \sum_{i_1 \cdots i_p} \Phi\,(v_{i_1}, \ldots, v_{i_p})$$

where $\Phi \in \otimes^p V^*$, $v_i \in V$ and sign $(i_1, \cdots, i_p)$ is the sign of the permutation $(i_1, \ldots, i_p)$ with respect to $(1, \ldots, p)$.

Then A, S have the same properties as before and $\otimes^p V^* = A\otimes^p V^* \oplus S \otimes^p V^*$.

12.2 DEFINITION *A tensor $\Phi \in \otimes^p V^*$ is called antisymmetric if $A\Phi = \Phi$. Φ is called symmetric if $S\Phi = \Phi$.*

All this is, of course, valid also for $\otimes^p V$, the space of all contravariant tensors of order p on V.

13.1 Mixed tensors

A real valued function $\Phi(v_1, \ldots, v_p, v_1^*, \ldots, v_q^*)$ of p arguments $v_1, \ldots, v_p \in V$ and q arguments $v_1^*, \ldots, v_q^* \in V^*$, which is linear in each of its arguments, is a tensor on V, covariant of order p and contravariant of order q. Such a tensor is called a *mixed tensor of type* (p, q) *on* V.

With the same operations as in 11.1, the set of all mixed tensors of type (p, q) on V is a vector space denoted by $(\otimes^p V^*) \otimes (\otimes^q V)$.

As an example, if $p = q = 1$, then $V^* \otimes V$ is a vector space of dimension n^2. A basis of $V^* \otimes V$ is given by the elements $b^{*i} \otimes b_j$, $(i, j = 1, \ldots, n)$ with

$$b^{*i} \otimes b_j (v, v^*) = \langle v, b^{*i}\rangle \langle b_j, v^*\rangle.$$

It follows that, for every $\Phi \in V^* \otimes V$, one has:

$$\Phi = \sum_{i,j=1}^{n} \Phi_i^j b^{*i} \otimes b_j,$$

where $\Phi_i^j = \Phi(b_i, b^{*j})$.

The components of Φ, with respect to a change of basis of V, transform with the law:

$$\Phi_{2i}^{j} = \sum_{h,k=1}^{n} a_h^j a_i^k \Phi_{1k}^{h}.$$

We remark that if $T: V \to V$ is a linear map, then

$$\langle Tv, v^*\rangle = \Phi(v, v^*),$$

$(v \in V,\ v^* \in V^*)$, is a mixed tensor of type $(1, 1)$ on V, whose components Φ_i^j in the basis $\{b^{*i} \otimes b_j\}$ of $V^* \otimes V$ are the entries of the matrix (a_i^j) which represents the linear map T in the basis $\{b_i\}$.

The generalization to the case (p, q), $p, q > 1$, is left to the reader.

14.1 Duality

The space of covariant tensors of order p on a finite dimensional vector space V is isomorphic to the space of contravariant tensors of order p on V since they are vector spaces of the same finite dimension. Moreover, if $\{b_i\}$ is a basis of V and $\{b^{*j}\}$ is the dual basis of V^*, we have:

$$\begin{aligned}\langle b_{i_1} \otimes \cdots \otimes b_{i_p}, b^{*j_1} \otimes \cdots \otimes b^{*j_p}\rangle &= \langle b_{i_1}, b^{*j_1}\rangle \cdots \langle b_{i_p}, b^{*j_p}\rangle \\ &= \delta_{i_1}^{j_1} \cdots \delta_{i_p}^{j_p},\end{aligned}$$

which shows that $\{b_{i_1} \otimes \cdots \otimes b_{i_p}\}$ and $\{b^{*j_1} \otimes \cdots \otimes b^{*j_p}\}$ are dual bases of $\otimes^p V$ and $\otimes^p V^*$.

We also have:

14.2 THEOREM *The space $\otimes^p V^*$ is isomorphic to the dual of $\otimes^p V$.* ○

It is interesting to construct such an isomorphism. Let $v_{i_1}, \ldots, v_{i_p} \in V$ so that $\{v_{i_1} \otimes \cdots \otimes v_{i_p}\}$ is a basis of $\otimes^p V$. Then we have:

$$v_{i_1} \otimes \cdots \otimes v_{i_p} (b^{*j_1}, \ldots, b^{*j_p}) = \langle v_{i_1}, b^{*j_1} \rangle \cdots \langle v_{i_p}, b^{*j_p} \rangle = v_{i_1}^{j_1} \cdots v_{i_p}^{j_p}.$$

Let $\psi \in \otimes^p V$, then

$$\psi = \sum_{i_1 \cdots i_p} \psi^{i_1 \cdots i_p} v_{i_1} \otimes \cdots \otimes v_{i_p}$$

and

$$\psi (b^{*j_1}, \ldots, b^{*j_p}) = \sum_{i_1 \cdots i_p} \psi^{i_1 \cdots i_p} v_{i_1}^{j_1} \cdots v_{i_p}^{j_p}.$$

Let $\Phi \in \otimes^p V^*$; define $\Phi (v_{i_1} \otimes \cdots \otimes v_{i_p}) = \Phi (v_{i_1}, \ldots, v_{i_p})$. Then the linear functional on $\otimes^p V$:

$$L(\psi) = \sum_{i_1 \cdots i_p} \psi^{i_1 \cdots i_p} \Phi (v_{i_1} \otimes \cdots \otimes v_{i_p}),$$

defines a map $\otimes^p V^* \to (\otimes^p V)^*$ which is clearly injective and therefore is an isomorphism.

15.1 Inner products

Let $\Phi \in \otimes^2 V^*$. Whenever we fix $v \in V$, the map: $w \to \Phi (v, w)$ is an element of V^*. Thus Φ defines a map $\bar{\Phi}$: $V \to V^*$ which to $v \in V$ associates $\bar{\Phi}(v) = \Phi (v, w)$. Conversely, if $\bar{\Phi}$ is any map $V \to V^*$, then $\Phi (v, w) = (\bar{\Phi}(v)) (w)$ is a covariant tensor of order two on V. The tensor Φ is said to be *non-degenerate* if the map $\bar{\Phi}$ is injective, that is, if $(\bar{\Phi}(v)) (w) = 0$, for all $w \in V$, implies $v = 0$.

If Φ is a non-degenerate tensor of order two, we can consider the map $\bar{\Phi}^{-1}$: $V^* \to V$. Then, just as before, we can associate to $\bar{\Phi}^{-1}$ a contravariant tensor of order two on V.

In other words, every non-degenerate covariant (contravariant) tensor of order two on V defines an isomorphism of V onto V^*. In the same way, every non-degenerate mixed tensor of order two on V defines an automorphism of V (or V^*).

15.2 DEFINITION *An inner product on V is any covariant tensor Φ, of order two on V, such that*

i) $\Phi(v,w) = \Phi(w, v)$ *for all* $v, w \in V$.
ii) $\Phi(v, v) > 0$ *if* $v \neq 0$.
iii) $\Phi(v, v) = 0$ *if and only if* $v = 0$.

If V is a complex vector space, in order that condition (ii) may make sense, condition (i) is replaced by:

$$\text{i}') \quad \Phi(v, w) = \overline{\Phi(w, v)},$$

where $\overline{\Phi(w, v)}$ denotes the conjugate of $\Phi(w, v)$.

A vector space V together with an inner product Φ on V is called an *inner product space*. If V is an inner product space, the notion of orthogonality of vectors can be defined. Two vectors $v, w \in V$ are said to be orthogonal if $\Phi(v, w) = 0$.

15.3 DEFINITION *Let W be a subspace of the inner product space V. The set,*

$$W^{\perp} = \{v \in V \mid \Phi(v, w) = 0 \quad \textit{for all} \quad w \in W\},$$

is called the orthogonal complement of W in V.

It is easy to show that $W^{\perp}$ is a subspace of V and that $W \cap W^{\perp} = 0 \in V$.

If V is an inner product space, for every $v \in V$, the *norm* of v, denoted by $\|v\|$, is defined by $\|v\| = \sqrt{\Phi(v, v)}$. We leave it to the reader to verify that $\|\cdot\|$ is actually a norm.

15.4 DEFINITION *A set $\{v_1, \ldots, v_k\}$ of vectors in an inner product space V is called an orthonormal set if:*

i) $\Phi(v_i, v_i) = 1$ *for every* i,
ii) $\Phi(v_i, v_j) = 0$ *if* $i \neq j$.

Vectors of an orthonormal set are clearly linearly independent. The proof of the following propositions is left to the reader. Let V be any (finite dimensional) inner product space; then:

15.5 PROPOSITION *V has a basis which is an orthonormal set.* ○

15.6 PROPOSITION *Let W be a subspace of V. Then $V = W \oplus W^{\perp}$ and $(W^{\perp})^{\perp} = W$.* ○

16.1 Grassmann algebra

Let V be an n-dimensional vector space. We have already considered the vector space $\otimes^p V$ of contravariant tensors of order p on V and the subspace $A \otimes {}^p V$ of antisymmetric contravariant tensors of order p on V.

$A\otimes^p V$ will be denoted by $\Lambda^p V$ and called the *p-fold exterior power* of V. Then we have $\Lambda^1 V = V$ and put $\Lambda^0 V = \mathbf{R}$.

If $\{b_i\}_{i=1,\dots,n}$ is a basis of V, we have seen that $\{b_{i_1}\otimes \cdots \otimes b_{i_p}\}$ is a basis of $\otimes^p V$. We now put:

$$b_{i_1}\wedge \cdots \wedge b_{i_p} = A\,(b_{i_1}\otimes \cdots \otimes b_{i_p}),$$

and define a bilinear map:

$$\Lambda^p V \times \Lambda^q V \to \Lambda^{p+q} V,$$

by sending $(b_{i_1}\wedge \cdots \wedge b_{i_p}, b_{j_1}\wedge \cdots \wedge b_{j_q})$ into $b_{i_1}\wedge \cdots \wedge b_{i_p}\wedge b_{j_1}\wedge \cdots \wedge b_{j_q}$.

We have:

$$b_{j_1}\wedge \cdots \wedge b_{j_q}\wedge b_{i_1}\wedge \cdots \wedge b_{i_p} = (-1)^{pq}\, b_{i_1}\wedge \cdots \wedge b_{i_p}\wedge b_{j_1}\wedge \cdots \wedge b_{j_q}.$$

Then, for $\Phi \in \Lambda^p V$ and $\psi \in \Lambda^q V$, the *exterior product* $\Phi \wedge \psi$ is defined by linearity and we have:

16.2 THEOREM *If $\Phi \in \Lambda^p V$ and $\psi \in \Lambda^q V$, then*

$$\psi \wedge \Phi = (-1)^{pq} \Phi \wedge \psi. \quad \bigcirc$$

16.3 COROLLARY *If $p > n$, then $\Lambda^p V = 0$.* ○

From the above theorem it follows that the $\binom{n}{p}$ elements

$$\{b_{i_1}\wedge \cdots \wedge b_{i_p}\}_{1 \le i_1 < \cdots < i_p \le n}$$

form a basis of the vector space $\Lambda^p V$. Therefore, if $\Phi \in \Lambda^p V$, we have:

$$\Phi = \sum_{1 \le i_1 < \cdots < i_p \le n} \Phi^{i_1 \cdots i_p} b_{i_1}\wedge \cdots \wedge b_{i_p}.$$

We consider now the direct sum of the vector spaces $\Lambda^p V$ for $p = 0, 1, \dots, n$ and put:

$$\Lambda V = \Lambda^0 V \oplus \Lambda^1 V \oplus \cdots \oplus \Lambda^n V.$$

An element $\Phi \in \Lambda V$ is written in a unique way as:

$$\Phi = \Phi^0 + \Phi^1 + \cdots + \Phi^n,$$

with $\Phi^i \in \Lambda^i V$.

If $\Phi, \psi \in \wedge V$, then a multiplication is defined in $\wedge V$ by:

$$\Phi\psi = \sum_{i=0}^{n} \sum_{j=0}^{i} \Phi^j \wedge \psi^{i-j},$$

where $\Phi^0 \wedge \psi^i = \Phi^0\psi^i$ and $\Phi^i \wedge \psi^0 = \Phi^i\psi^0$.

Thus $\wedge V$ becomes an algebra of dimension 2^n called the *Grassmann algebra* of V.

17.1 Problems

1 (1.1) Prove lemmas 1.7 and 1.8.

2 (2.1) Fill in all the missing details in examples 2.1.

3 (3.1) Let V, W be two finite dimensional vector spaces. Prove that $\dim \operatorname{Hom}(V, W) = \dim V \cdot \dim W$.

4 (3.1) Let V, W be finite dimensional vector spaces and let $T \in \operatorname{Hom}(V, W)$. Prove that:

$$\dim V = \dim(\operatorname{Ker} T) + \dim(TV).$$

5 (7.1) Prove that the vector space $H_0(X, \mathbf{R})$ has the following universal property: given any vector space V there exists a set X such that $V \simeq H_0(X, \mathbf{R})$.

6 (9.1) Define the direct sum $\bigoplus_{i=1}^{n} V_i$ of a finite number of vector spaces $V_1, \ldots, V_n$.

7 (9.1) Let W be a subspace of a finite dimensional vector space V. Prove that:

$$\dim(V/W) = \dim V - \dim W.$$

8 (9.1) Let W, Z be two subspaces of a finite dimensional vector space V. The set $\{w + z \mid w \in W, z \in Z\}$ is a subspace of V called the internal direct sum of W and Z. Prove that.

a) the internal direct sum is isomorphic to the external direct sum $W \oplus Z$,

b) V is the internal direct sum of W and Z if and only if $W \cap Z = 0$ and $W \cup Z = V$.

9 (10.1) Given vector spaces $V_1, \ldots, V_n$ define the tensor product $\bigotimes_{i=1}^{n} V_i$.

10 (10.1) Let V, W be vector spaces. Prove that $V \otimes W \simeq W \otimes V$.

11 (10.1) Let V, W, Z be vector spaces. Prove that:

$$(V \otimes W) \otimes Z \simeq V \otimes (W \otimes Z).$$

12 (10.1) If V is a vector space, prove that $V \simeq V \otimes \mathbf{R}$.

13 (10.1) Let V, W, Z be vector spaces. Prove that

$$(V \oplus W) \otimes Z \simeq (V \otimes Z) \oplus (W \otimes Z).$$

14 (11.1) Generalize theorem 10.4 to the case of p-linear maps; that is, prove the existence of an isomorphism:

$$\text{Hom}\,(V_1, \ldots, V_n; M) \simeq \text{Hom}\,(V_1 \otimes \cdots \otimes V_n, M).$$

15 (15.1) Prove propositions 15.5 and 15.6.

16 (16.1) Prove theorem 16.2.

17 (16.1) If V, W are vector spaces and A is the map defined in section 12.1, prove the existence of an isomorphism:

$$\text{Hom}\,(A \otimes^2 V, W) \simeq \text{Hom}\,(\wedge^2 V, W).$$

Generalize to:

$$\text{Hom}\,(A \otimes^p V, W) \simeq \text{Hom}\,(\wedge^p V, W).$$

18 (16.1) Let $v_1, \ldots, v_p$ be elements of a vector space V. Prove that $v_1, \ldots, v_p$ are linearly independent if and only if $v_1 \wedge \cdots \wedge v_p \neq 0$.

19 (16.1) Prove that, if V is an n-dimensional vector space, then the Grassmann algebra of V has dimension 2^n.

CHAPTER 2

Euclidean spaces

1.1 Structures in $\mathbf{R}^n$

THE SET $\mathbf{R}^n$ of all ordered n-tuples $x = \{x_1, ..., x_n\}$ of real numbers was made into a vector space by the operations:

$$x + y = \{x_1, ..., x_n\} + \{y_1, ..., y_n\} = \{x_1 + y_1, ..., x_n + y_n\},$$
$$cx = c\{x_1, ..., x_n\} = \{cx_1, ..., cx_n\},$$

and denoted by V^n.

V^n, with the topology of the n-fold topological product of the real line with itself, becomes a topological vector space in the sense that the maps:

$$\mathbf{R}^n \times \mathbf{R}^n \to \mathbf{R}^n$$

defined by $(x, y) \to (x + y)$ and

$$\mathbf{R} \times \mathbf{R}^n \to \mathbf{R}^n$$

defined by $(c, x) \to cx$ are continuous with respect to that topology.

$\mathbf{R}^n$ can also be made into a metric space by introducing a metric given by the distance function:

$$d(x, y) = [(x_1 - y_1)^2 + \cdots + (x_n - y_n)^2]^{1/2},$$

$x, y \in \mathbf{R}^n$. Considered as a metric space, $\mathbf{R}^n$ is called the *n-dimensional Euclidean space* and is denoted by E^n.

We remark that the topology induced in E^n by the metric is equivalent to that which we have already put on V^n.

When $\mathbf{R}^n$ is endowed simultaneously with all these structures, the check of the following properties is straightforward:

i) every isometry of E^n, that is, every map $\varrho: E^n \to E^n$ such that $d(x, y) = d(\varrho x, \varrho y)$ for all $x, y \in \mathbf{R}^n$ is a homeomorphism. Isometries of E^n form a group, the group of *rigid motions*, which is denoted by $\mathscr{R}(n)$;

ii) every element $T \in \text{Hom}(V^n, V^n)$ is continuous and, in particular, every element $S \in GL(n, \mathbf{R})$ is a homeomorphism of E^n onto itself;

iii) the additive group of V^n acts on E^n as a group of homeomorphisms.

2.1 Differentiable functions

We shall consider the topological space E^n together with its vector space structure of V^n. We shall assume, moreover, that the canonical basis B_0 (1.7.3) is fixed in V^n. Then an n-tuple $\{x_1, \ldots, x_n\} \in \mathbf{R}^n$ may be identified with the point $x \in E^n$, which it defines, and with the vector of V^n, whose components with respect to the basis B_0 are $x_1, \ldots, x_n$. If $x = \{x_1, \ldots, x_n\}$ is a point of E^n, the real numbers $x_1, \ldots, x_n$ are called the *coordinates* of x.

2.2 DEFINITION *A real valued function f, defined on an open set $U \subset E^n$, is called differentiable on U if f has partial derivatives of all orders with respect to the coordinates $x_1, \ldots, x_n$ at every point $x \in U$.*

A differentiable function, which at every point of its domain of definition U can be expanded into a convergent power series, with positive radius of convergence, is called *analytic on U*.

A differentiable function on U is also called $C^\infty(U)$. We shall simply write C^∞ instead of $C^\infty(U)$ when no confusion is possible. The set of all differentiable functions on an open set $U \subset E^n$ is a ring under pointwise addition and multiplication of functions. We shall denote this ring by $\mathscr{D}(U)$.

2.3 DEFINITION *Let U be an open subset of E^n. A map $f\colon U \to E^m$ is called a differentiable map if, for every open set $W \subset E^m$ and differentiable function g on W, the composite function $g \circ f$ is differentiable on $f^{-1}(W)$.*

It follows from the definition that a map $f\colon U \to E^m$ is a differentiable map if and only if the coordinates $y_1, \ldots, y_m$ of points $f(x) \in E^m$ are differentiable functions on U of the coordinates $x_1, \ldots, x_n$ of points $x \in U$. Also, since a map $f\colon U \to E^m$ is given by m real valued functions $f_1, \ldots, f_m$, then f is a differentiable map if and only if $f_1(x), \ldots, f_m(x)$ are differentiable functions on U.

Let U, W be open subsets of E^n. A map $f\colon U \to W$ is called *injective* if f is one to one and *surjective* if $f(U) = W$. A map $f\colon U \to W$ which is both injective and surjective is called *bijective*.

2.4 DEFINITION *Let U, W be open subsets of E^n. A bijective map $f\colon U \to W$, is called a diffeomorphism if f is differentiable on U and f^{-1} is differentiable on W.*

That, even if f is a differentiable homeomorphism, f need not be a diffeomorphism is shown by the following example. For $k > 1$, k odd integer, the map $x^k: \mathbf{R} \to \mathbf{R}$ is a differentiable homeomorphism, but the inverse map is not differentiable on any neighborhood of the origin.

We shall show now that for every point $x \in E^n$ we can give a characterization of the vector space V^n in terms of the ring $\mathscr{D}(U)$ of real valued functions which are C^∞ on some open neighborhood U of x. Addition of functions and multiplication by scalars make $\mathscr{D}(U)$ into a vector space, as it was shown in example 1.7.4.

It is easily seen that $\mathscr{D}(U)$ is an infinite dimensional vector space.

Let $x = \{x_1, \ldots, x_n\}$ be a fixed point of E^n, let U be any open neighborhood of x and let $v = (v^1, \ldots, v^n)$ be a vector of V^n. For every $f \in \mathscr{D}(U)$ we consider the function $f(x + \tau v)$, $\tau \in \mathbf{R}$, and put

$$f_v(x) = \frac{d}{d\tau} f(x + \tau v)\bigg|_{\tau=0} .$$

If $v = e_i$, then we have,

$$f_{e_i}(x) = \frac{\partial f}{\partial x_i}(x);$$

and, in general:

$$f_v(x) = \sum_{i=1}^{n} v^i \frac{\partial f}{\partial x_i}(x).$$

Therefore, to every vector $v \in V^n$, we can associate a map $L_{x,v}: \mathscr{D}(U) \to \mathbf{R}$ by putting:

$$L_{x,v}(f) = f_v(x).$$

It is easy to see that the map $L_{x,v}$ satisfies the following properties:

2.5

i) $L_{x,v}(f + g) = L_{x,v}(f) + L_{x,v}(g)$,

ii) $L_{x,v}(fg) = f(x) L_{x,v}(g) + g(x) L_{x,v}(f)$,

iii) $L_{x,v}(f) = 0$ if $f =$ constant.

It follows from (ii) and (iii) that, for $c \in \mathbf{R}$, one has:

$$L_{x,v}(cf) = cL_{x,v}(f).$$

Therefore, $L_{x,v}$ is a linear functional on $\mathscr{D}(U)$. This functional is called a *derivation* on the ring $\mathscr{D}(U)$.

An important consequence of property (iii) is that $L_{x,v}(f) = L_{x,v}(g)$ if f and g differ by a constant on some neighborhood of x.

Now if $v, w \in V^n$ and $c \in \mathbf{R}$, an elementary computation shows that:

$$L_{x,v}(f) + L_{x,w}(f) = L_{x,v+w}(f),$$

and

$$L_{x,cv}(f) = cL_{x,v}(f),$$

for every $f \in \mathscr{D}(U)$. Thus, for every fixed $x \in U$, the set of all linear functionals $L_{x,v}$ is a vector space.

Let $\pi_i \in \mathscr{D}(U)$ be the function $\pi_i(x_1, \ldots, x_n) = x_i$. Then clearly $L_{x,v}(\pi_i) = v^i$ for every $v \in V^n$. It follows that, if $L_{x,v}(f) = L_{x,w}(f)$ for all $f \in \mathscr{D}(U)$, then $v = w$. In other words, the map $v \to L_{x,v}$ is injective.

We shall show now that if for fixed x, l is a derivation on $\mathscr{D}(U)$, that is, l is a functional on $\mathscr{D}(U)$ such that:

i) $l(f + g) = l(f) + l(g)$,

2.6 ii) $l(fg) = f(x)\, l(g) + g(x)\, l(f)$,

iii) $l(f) = 0$ if $f =$ constant,

then there exists a vector $v \in V^n$ such that, for every $f \in \mathscr{D}(U)$, $l(f) = L_{x,v}(f)$.

We need the following lemma:

2.7 LEMMA *Let $f \in \mathscr{D}(U)$. Then there exist functions $g_i \in \mathscr{D}(U)$ such that in a suitable neighborhood of x:*

$$f = f(x) + \sum_{i=1}^{n} (\pi_i - x_i)\, g_i .$$

Moreover,

$$g_i(x) = \frac{\partial f}{\partial x_i}(x).$$

PROOF Let $W \subset U$ be a convex neighborhood of x and let $y \in W$. Consider the function:

$$\alpha(\tau) = f[x_1 + \tau(y_1 - x_1), \ldots, x_n + \tau(y_n - x_n)]$$

for $0 \leq \tau \leq 1$ and write

$$\alpha(1) - \alpha(0) = \int_0^1 \alpha'(\tau)\, d\tau.$$

Then

$$f(y) = f(x) + \sum_{i=1}^{n} (y_i - x_i)\, g_i(y)$$

where

$$g_i(y) = \int_0^1 \frac{\partial f}{\partial x_i} [x_1 + \tau (y_1 - x_1), \ldots, x_n + \tau (y_n - x_n)] \, d\tau. \quad \bigcirc$$

We now use lemma 2.7 to compute $l(f)$. We have:

2.8
$$l(f) = \sum_{i=1}^{n} l [(\pi_i - x_i) g_i] = \sum_{i=1}^{n} l(\pi_i) g_i(x)$$
$$= \sum_{i=1}^{n} l(\pi_i) \frac{\partial f}{\partial x_i} (x_1, \ldots, x_n).$$

Therefore we have $l(f) = L_{x,v}(f)$ with $v = \{l(\pi_1), \ldots, l(\pi_n)\}$. Thus the map $v \to L_{x,v}$ is surjective.

3.1 Tangent space at a point E^n

Let $x \in E^n$ and let U be an open neighborhood of x. We shall denote by $T_x(E^n)$ the set of all functionals on $\mathscr{D}(U)$ satisfying conditions 2.6. It is clear that the definition of $T_x(E^n)$ does not depend on U. We have already seen that $T_x(E^n)$ is a vector space. Moreover, $\dim T_x(E^n) = n$ since the map $v \to L_{x,v}$ is an injective map of V^n onto $T_x(E^n)$.

3.2 Definition *$T_x(E^n)$ is called the tangent space to E^n at x. Its elements are called tangent vectors to E^n at x.*

When no confusion is possible we shall simply write T_x instead of $T_x(E^n)$. Tangent vectors have an interesting geometrical interpretation which will be shown in 3.8.1.

That T_x is an n-dimensional vector space was proved by using both structures of E^n and V^n. We shall show now that the consideration of V^n can be avoided.

We remark first that $\mathscr{D}(U)$ is invariant under diffeomorphisms of U onto itself. In other words, diffeomorphisms $g: U \to U$ are characterized as those homeomorphisms of U onto itself such that $f \in \mathscr{D}(U)$ if and only if $f \circ g \in \mathscr{D}(U)$.

If $x \in U$ and $g = (g_1, \ldots, g_n) : U \to U$ is a diffeomorphism, then $g_1(x), \ldots, g_n(x)$ are coordinates of x and g is said to be *a coordinate system* on U. In particular, the identity $i: U \to U$ is that coordinate system which to the point $x = \{x_1, \ldots, x_n\}$ associates the coordinates $x_1, \ldots, x_n$. In what follows, the coordinates of $x \in E^n$ will always be denoted by $x_1, \ldots, x_n$ and meant up to diffeomorphisms of a neighborhood of x.

The tangent space T_x at $x \in E^n$ is defined as the set of all functionals t: $\mathscr{D}(U) \to \mathbf{R}$ which satisfy conditions 2.6. By definition T_x is a vector space. In order to show that $\dim T_x = n$, it is enough to show that $\left\{\frac{\partial}{\partial x_1}, \ldots, \frac{\partial}{\partial x_n}\right\}$ is a basis of T_x. This follows at once from 2.8; in fact, for every $t \in T_x$, we have

$$t = \sum_{i=1}^{n} t(\pi_i) \frac{\partial}{\partial x_i}.$$

3.3 *Remark* The linear functional t takes the same value at functions which coincide on a neighborhood of x. Thus for the consideration of tangent vectors, it should be possible to identify two functions if they coincide on some neighborhood of x.

Moreover, tangent vectors at x do not depend on the neighborhood U containing x.

We shall denote by $\mathscr{D}_x$ the set of all real valued functions differentiable on some neighborhood of $x \in E^n$. If $f, g \in \mathscr{D}_x$, then $f + g$ and fg are also in $\mathscr{D}_x$ since they are differentiable on the intersection of the two neighborhoods of x where f, g are differentiable respectively. Although it might be tempting to use $\mathscr{D}_x$ in order to define tangent vectors at x, some other considerations are required because $\mathscr{D}_x$ does not have nice algebraic structure. This will be done in section 3.6.1.

4.1 Cotangent space at a point of E^n

T_x is a vector space of finite dimension n; we shall denote by T_x^* its dual space. T_x^* is also an n-dimensional vector space.

4.2 DEFINITION *T_x^* is called the cotangent space to E^n at x. The elements of T_x^* are called cotangent vectors to E^n at x.*

T_x^* is isomorphic to T_x; however, there is not a natural isomorphism of T_x onto T_x^*. In fact, every basis $B = \{b_i\}$ of T_x determines canonically the dual basis $B^* = \{b^{*i}\}$ of T_x^* and therefore the isomorphism α_{B^*B}: $T_x \to T_x^*$ under which the image of the tangent vector $t = \sum_{i=1}^{n} t^i b_i$ is the cotangent vector $t^* = \sum_{i=1}^{n} t_i^* b^{*i}$ with $t_i^* = t^i$, $(i = 1, \ldots, n)$. But the isomorphism α_{B^*B} depends on the choice of the basis B.

Every system of coordinates on a neighborhood of x induces a basis $\left\{\frac{\partial}{\partial x_i}\right\}$ of T_x and these systems of coordinates are always defined up to a diffeomorphism of a neighborhood of x.

Let $x \in E^n$, let U be an open neighborhood of x, let $t \in T_x$, and let $f \in \mathscr{D}(U)$. The operation $f_t(x)$ which we defined in section 2.1 may be considered in two different ways, namely,

i) for fixed t as a functional on $\mathscr{D}(U)$,
ii) for fixed f as a functional on T_x.

We have seen that (i) leads to the definition of tangent vectors. We shall show now that (ii) leads to the notion of cotangent vectors.

Let d: $\mathscr{D}(U) \to T_x^*$ be the map which to the function $f \in \mathscr{D}(U)$ associates the vector $df \in T_x^*$, defined by:

$$\langle t, df \rangle = \langle f, t \rangle, \; t \in T_x.$$

In particular, we have:

$$\left\langle \frac{\partial}{\partial x_i}, d\pi_j \right\rangle = \delta_i^j,$$

which shows that $\{d\pi_j\}$ is the basis of T_x^* dual of the basis $\left\{\frac{\partial}{\partial x_i}\right\}$ of T_x.

We shall write dx^i for $d\pi_i$.

Then, for every $f \in \mathscr{D}(U)$, we have:

$$df = \sum_{i=1}^{n} (df)_i \, dx^i,$$

with:

$$(df)_i = \left\langle \frac{\partial}{\partial x_i}, df \right\rangle = \frac{\partial f}{\partial x_i},$$

and thus:

$$df = \sum_{i=1}^{n} \frac{\partial f}{\partial x_i} \, dx^i.$$

The map d we have just defined is called *differential*. Thus, cotangent vectors at x may be identified with differentials of functions at x.

4.3 *Remark* The cotangent vector df depends only on the values of f on a neighborhood of x. Therefore, $df = dg$ if f and g coincide, or differ by a constant, on some neighborhood of x.

4.4 *Remark* Linear maps $T_x \to T_x^*$ may be expressed by covariant tensors of order two on T_x as it was shown in 1.13.1. In particular, every isomorphism of T_x onto T_x^* is expressed by a non-degenerate covariant tensor of order two on T_x.

4.5 *Remark* The tangent and cotangent spaces T_x, T_x^* at every point $x \in E^n$ are both isomorphic to V^n. Therefore the definition of tangent and cotangent vectors assigns to every $x \in E^n$ two copies of V^n. There are actually three copies of V^n since, for fixed x, we can disregard the other structures of E^n. In this situation, we can identify these spaces and consider the elements of T_x and T_x^* as belonging to E^n. Examples from elementary geometry, like the tangent plane at a point of a surface in E^3, offer more interesting situations. In the next chapter, we shall show how two dual vector spaces can be assigned to a point not belonging to a linear space.

5.1 Contravariant vector fields

Let U be an open subset of E^n. *A contravariant vector field* $L = \{L_x\}$ *on* U is a function which to every point $x \in U$ assigns a vector $L_x \in T_x$.

Let L be a contravariant vector field on U and $f \in \mathscr{D}(U)$. If $L_x \in T_x$ is the vector which L assigns to the point $x \in U$, then to f is associated a real valued function Lf defined on U by:

$$Lf(x) = \langle f, L_x \rangle.$$

5.2 Definition *A contravariant vector field L is said to be* $C^\infty(U)$ *if* $Lf \in \mathscr{D}(U)$ *for every* $f \in \mathscr{D}(U)$.

We remark that if, for every $x \in U$, we choose in T_x the basis $\left\{\dfrac{\partial}{\partial x_i}\right\}$, then the assignment of a contravariant vector field L is equivalent to the assignment of n real valued functions L^i on U, and we have:

$$Lf(x) = \sum_{i=1}^{n} L^i(x) \frac{\partial f}{\partial x_i}(x),$$

for every $f \in \mathscr{D}(U)$.

It follows that if the n functions L^i are C^∞, then L is a C^∞ vector field. Conversely, if L is $C^\infty(U)$, by taking $f = \pi_i$, we get $L\pi_i = L^i$ and thus the L^i are C^∞. Therefore, a contravariant vector field L is C^∞ if and only if the n functions L^i are C^∞.

We have seen that a C^∞ contravariant vector field on U associates to every function $f \in \mathscr{D}(U)$ a new function $Lf \in \mathscr{D}(U)$ and thus defines a map L: $\mathscr{D}(U) \to \mathscr{D}(U)$. It is easily seen that L satisfies the properties:

5.3

i) $L(f + g) = L(f) + L(g)$

ii) $L(fg) = fL(g) + gL(f)$

iii) $L(f) = 0$ if $f =$ constant.

We shall therefore say that L is a *derivation* on $\mathscr{D}(U)$.

We proved that analogous properties 2.6 characterize tangent vectors. We shall show now that 5.3 characterize contravariant vector fields. In fact, let L: $\mathscr{D}(U) \to \mathscr{D}(U)$ be a derivation and, for every $x \in U$, let $t_x \in T_x$ be a vector such that $\langle f, t_x \rangle = Lf(x)$. Then the contravariant vector field $T = \{t_x\}$ induces the derivation L.

Let $\mathscr{V}(U)$ be the set of all C^∞ contravariant vector fields on an open subset U of E^n. For $L, M \in \mathscr{V}(U)$ and $f \in \mathscr{D}(U)$, the sum $L + M$ and the product fL are defined by:

$$(L + M)_x = L_x + M_x,$$

$$(fL)_x = f(x) L_x,$$

for $x \in U$.

Since, clearly, $L + M$ and fL are in $\mathscr{V}(U)$, these operations make $\mathscr{V}(U)$ into a module over the ring $\mathscr{D}(U)$.

For every $f \in \mathscr{D}(U)$ and $L, M \in \mathscr{V}(U)$, we put:

$$[L, M](f) = L(Mf) - M(Lf).$$

The map $[L, M]$: $\mathscr{D}(U) \to \mathscr{D}(U)$ so obtained is a derivation on $\mathscr{D}(U)$, as it is easy to verify. The contravariant vector field associated to this derivation is called the *bracket* of L and M and will also be denoted by $[L, M]$.

The bracket operation satisfies the following properties:

i) $[L, M] = -[M, L]$,

ii) $[L, [M, N]] + [M, [N, L]] + [N, [L, M]] = 0$,

for $L, M, N \in \mathscr{V}(U)$. Property (ii) is called *Jacobi identity*.

The set $\mathscr{V}(U)$ of all C^∞ contravariant vector fields on U, with the above operations, is a Lie algebra.

6.1 Covariant vector fields

A covariant vector field $L^* = \{L_x^*\}$ on an open subset U of E^n is a function which to every point $x \in U$ assigns a vector $L_x^* \in T_x^*$.

A covariant vector field L^* on U associates to every contravariant vector field M on U the function $L^*(M)$ defined on U by:

$$L^*(M)(x) = \langle M_x, L_x^* \rangle.$$

6.2 DEFINITION *A covariant vector field L^* is said to be $C^\infty(U)$ if $L^*(M) \in \mathscr{D}(U)$ whenever M is a C^∞ contravariant vector field.*

If, for every $x \in U$ we choose in T_x^* the basis $\{dx^i\}$, then the assignment of a covariant vector field L^* is equivalent to the assignment of n real valued functions L_i^* on U, and

$$L^*(x) = \sum_{i=1}^{n} L_i^*(x)\, dx^i.$$

Then a vector field L^* is C^∞ if and only if the L_i^* are C^∞ functions.

6.3 *Example* Let $f \in \mathscr{D}(U)$. If to every $x \in U$ we associate $df(x)$, we get a covariant vector field on U which is denoted by df and called the *differential* of f. df is a C^∞ vector field, since, for every C^∞ contravariant vector field L on U, we have $df(L) = Lf$. In fact at every $x \in U$, $df(L)(x) = \langle L_x, df(x) \rangle = \langle f, L_x \rangle = Lf(x)$.

It is well-known that not every covariant field is of type df. A covariant vector field L^* is called *exact* if there exists a function f such that $L^* = df$.

7.1 Tensor fields

A tensor field $\Phi = \{\Phi_x\}$, covariant of order p and contravariant of order q, on a subset U of E^n is a function which, to every $x \in U$, assigns a tensor Φ_x, covariant of order p and contravariant of order q.

A tensor field Φ is said to be $C^\infty(U)$ if, for every p-tuple $L_1, \ldots, L_p$ of C^∞ contravariant vector fields on U and every q-tuple $L_1^*, \ldots, L_q^*$ of C^∞ covariant vector fields on U, the function $\Phi(L_1, \ldots, L_p, L_1^*, \ldots, L_q^*)$, defined on U by:

$$\Phi(L_1, \ldots, L_p, L_1^*, \ldots, L_q^*)(x) = \Phi_x(L_{1x}, \ldots, L_{px}, L_{1x}^*, \ldots, L_{qx}^*)$$

is $C^\infty(U)$.

If $\{b_i\}$, $\{b^{*j}\}$ are dual bases of T_x, T_x^*, then:

$$\Phi_x = \sum_{\substack{i_1 \cdots i_p \\ j_1 \cdots j_q}} \Phi_{x,\, i_1 \cdots i_p}^{\ j_1 \cdots j_q} b^{*i_1} \otimes \cdots \otimes b^{*i_p} \otimes b_{j_1} \otimes \cdots \otimes b_{j_q}$$

with

$$\Phi_{x,\, i_1 \cdots i_p}^{\ j_1 \cdots j_q} = \Phi_x (b_{i_1}, \ldots, b_{i_p}, b^{*j_1}, \ldots, b^{*j_q}).$$

In what follows we shall consider covariant tensor fields of order two on an open subset U of E^n.

Let Φ be such a field. For every $x \in U$ let the bases $\left\{\dfrac{\partial}{\partial x_i}\right\}$, $\{dx^i\}$ be fixed in T_x, T_x^*. Then we have:

$$\Phi_x = \sum_{i,j=1}^{n} \Phi_{ij}(x)\, dx^i \otimes dx^j,$$

where

$$\Phi_{ij}(x) = \Phi_{x,\, ij} = \Phi_x \left(\frac{\partial}{\partial x_i}, \frac{\partial}{\partial x_j}\right).$$

As we have remarked, Φ is C^∞ if and only if the functions Φ_{ij} are C^∞.

A covariant tensor field Φ will be called antisymmetric (symmetric) if Φ_x is an antisymmetric (symmetric) tensor for every $x \in U$.

The elements:

$$dx^i \wedge dx^j = A\,(dx^i \otimes dx^j) = \tfrac{1}{2}\,(dx^i \otimes dx^j - dx^j \otimes dx^i),$$

$$dx^i \vee dx^j = S\,(dx^i \otimes dx^j) = \tfrac{1}{2}\,(dx^i \otimes dx^j + dx^j \otimes dx^i),$$

form bases of $A \otimes^2 T_x^* = \Lambda^2 T_x^*$ and $S \otimes^2 T_x^*$ respectively.

In particular, if Φ is an antisymmetric covariant tensor field, then we have:

$$\Phi = \sum_{1 \le i < j \le n} \Phi_{ij}(x)\, dx^i \wedge dx^j.$$

Let Φ be a C^∞ covariant tensor field on U and let L, M be C^∞ contravariant vector fields on U. Consider the function:

$$(L, M) \to \Phi\,(L, M),$$

defined on $\mathscr{V}(U) \times \mathscr{V}(U)$. This function satisfies a linearity property with respect to $\mathscr{D}(U)$ in the sense that, for $L, M, L_1, L_2, M_1, M_2 \in \mathscr{V}(U)$ and $f \in \mathscr{D}(U)$, we have:

i) $\Phi\,(L_1 + L_2, M) = \Phi\,(L_1, M) + \Phi\,(L_2, M)$

7.2 ii) $\Phi\,(fL, M) = f\Phi\,(L, M)$

iii) $\Phi\,(L, M_1 + M_2) = \Phi\,(L, M_1) + \Phi\,(L, M_2)$

iv) $\Phi\,(L, fM) = f\Phi\,(L, M)$.

7.3 THEOREM *Let $\tilde{\Phi}$ be a C^∞ real valued function defined on $\mathscr{V}(U) \times \mathscr{V}(U)$ satisfying properties* 7.2. *Then there exists a unique C^∞ tensor field Φ on U such that $\tilde{\Phi} = \Phi$.*

Proof We prove first that if L is a C^∞ contravariant vector field on U such that $L_y = 0$ for every y in a neighborhood V of $x \in U$, then, for every contravariant vector field M, the function $\tilde{\Phi}(L, M)$ vanishes on a neighborhood $W \subset V$ of x.
Let $\eta \in \mathscr{D}(U)$ such that:

$$\eta(y) = \begin{cases} 1 & \text{if } \ y \in U \backslash V^* \\ 0 & \text{if } \ y \in W. \end{cases}$$

Then we have:

$$\tilde{\Phi}(L, M) = \tilde{\Phi}(\eta L, M) = \eta \tilde{\Phi}(L, M),$$

and thus:

$$\tilde{\Phi}(L, M)(y) = 0 \quad \text{if } y \in W.$$

Now, for each $x \in U$ and every pair s_x, t_x of vectors in T_x, we define a tensor Φ_x by:

$$\Phi_x(s_x, t_x) = \tilde{\Phi}(L, M)(x),$$

where L, M are two C^∞ vector fields such that $L_x = s_x$, $M_x = t_x$.

We must show that $\Phi_x(s_x, t_x)$ does not depend on the choice of L, M. Let $x_1, \ldots, x_n$ be coordinates of x. Then

$$L_x = \sum_{i=1}^{n} l^i(x) \frac{\partial}{\partial x_i}, \quad M_x = \sum_{i=1}^{n} m^i(x) \frac{\partial}{\partial x_i},$$

for $x \in U$.

Let L', M' be two other C^∞ vector fields such that $L'_x = s_x$, $M'_x = t_x$. Then:

$$L'_x = \sum_{i=1}^{n} l'^i(x) \frac{\partial}{\partial x_i}, \quad M'_x = \sum_{i=1}^{n} m'^i(x) \frac{\partial}{\partial x_i},$$

with $l'^i(x) = l^i(x)$, $m'^i(x) = m^i(x)$, $i = 1, \ldots, n$. Hence, for the value of $\tilde{\Phi}$ at x, we have

$$\tilde{\Phi}(L, M)(x) = \sum_{i,j=1}^{n} l^i(x)\, m^j(x)\, \tilde{\Phi}\left(\frac{\partial}{\partial x_i}, \frac{\partial}{\partial x_j}\right)$$

$$= \sum_{i,j=1}^{n} l'^i(x)\, m'^j(x)\, \tilde{\Phi}\left(\frac{\partial}{\partial x_i}, \frac{\partial}{\partial x_j}\right) = \tilde{\Phi}(L', M')(x). \ \bigcirc$$

---

* $U \backslash V$ means $U - (U \cap V)$.

8.1 Differential of a covariant vector field

To every covariant vector field on U, one can associate an antisymmetric covariant tensor field on U in the following way:

Let L^* be a C^∞ covariant vector field on U. Let L, M be two C^∞ contravariant vector fields on U. We define on U the function $\Phi(L, M)$ by

$$\Phi(L, M) = L(L^*(M)) - M(L^*(L)) - L^*([L, M]).$$

8.2 Lemma *The function $\Phi(L, M)$ defined above satisfies properties* 7.2.

Proof (i) and (iii) are obvious. We prove (ii) and the proof of (iv) is similar. Put:

$$\Phi_1(L, M) = L(L^*(M)) - M(L^*(L)),$$

$$\Phi_2(L, M) = -L^*([L, M]),$$

then:

$$\Phi(L, M) = \Phi_1(L, M) + \Phi_2(L, M).$$

For $f \in \mathscr{D}(U)$ we have

$$\Phi_1(fL, M) = f\Phi_1(L, M) - M(f)\, L^*(L).$$

From the identity

$$[fL, M] = f[L, M] - M(f)\, L$$

follows

$$\Phi_2(fL, M) = f\Phi_2(L, M) + M(f)\, L^*(L)$$

and thus

$$\Phi(fL, M) = f\Phi(L, M). \quad \bigcirc$$

Lemma 8.2 proves that Φ satisfies the hypotheses of theorem 7.3. Hence there exists a C^∞ covariant tensor field of order two on U, which we denote by dL^*, such that for the associated function dL^* we have $dL^* = \Phi$.

8.3 Definition *The tensor field dL^* is called the differential of the vector field L^*.*

Since $\Phi(L, M) = -\Phi(M, L)$, the tensor field dL^* is antisymmetric. Thus we have

$$dL^* = \sum_{1 \le i < j \le n} dL^*_{ij}\, dx^i \wedge dx^j,$$

and it is easy to show that if $L^* = \sum_{i=1}^{n} L_i^* dx^i$, then

$$dL_{ij}^* = \frac{\partial L_j^*}{\partial x_i} - \frac{\partial L_i^*}{\partial x_j}.$$

A covariant vector field L^* on U is called *closed* if $dL^* = 0$; that is, if

$$\frac{\partial L_j^*}{\partial x_i} = \frac{\partial L_i^*}{\partial x_j}$$

for every $x \in U$.

If $L^* = df$ for some $f \in \mathscr{D}(U)$, then L^* is closed. In fact, for every pair L, M of contravariant vector fields, we have

$$\begin{aligned} dL^*(L, M) &= L(L^*(M)) - M(L^*(L)) - L^*([L, M]) \\ &= L(df(M)) - M(df(L)) - df([L, M]) \\ &= L(M(f)) - M(L(f)) - [L, M](f) = 0. \end{aligned}$$

Actually, if $L^* = df$ for some $f \in \mathscr{D}(U)$, in a neighborhood of $x \in U$ we have $L^* = \sum_{i=1}^{n} \frac{\partial f}{\partial x_i} dx^i$. Hence,

$$dL_{ij}^* = \frac{\partial^2 f}{\partial x_i \, \partial x_j} - \frac{\partial^2 f}{\partial x_j \, \partial x_i} = 0.$$

We recall that a C^∞ vector field L^* is exact if $L^* = df$ for some $f \in \mathscr{D}(U)$. Therefore, every exact vector field is closed.

That not every closed vector field is exact follows from the fact that the conditions:

$$\frac{\partial L_j^*}{\partial x_i} = \frac{\partial L_i^*}{\partial x_j},$$

imply the existence of an $f \in \mathscr{D}(U)$ such that $L_i^* = \frac{\partial f}{\partial x_i}$ only locally. In general, there is not an $f \in \mathscr{D}(U)$ such that $L^* = df$.

9.1 Problems

1 (1.1) Prove statements (i), (ii), and (iii) of section 1.1.

2 (2.1) Let U be an open subset of E^n and let V be an open subset of E^m. Let $f : U \to V$ and $g : V \to E^r$ be differentiable maps. Prove that the composite map $g \circ f : U \to E^r$ is differentiable.

3 (2.1) Prove that for $v, w \in V^n$ and $f \in \mathscr{D}(U)$:

$$f_v(x) + f_w(x) = f_{v+w}(x),$$

and for $v \in V^n$, $c \in \mathbf{R}$ and $f \in \mathscr{D}(U)$:

$$f_{cv}(x) = cf_v\,(x).$$

4 (3.1) Let U be an open subset of E^n and let $x \in U$. Prove that a linear functional L on $\mathscr{D}(U)$ such that,

$$L\,(fg) = f(x)\,L(g) + g(x)\,L(f)$$

for every $f, g \in \mathscr{D}(U)$, satisfies also the property:

$$L(f) = 0 \text{ if } f = \text{constant}$$

and thus is a tangent vector at x.

5 (4.1) Prove that a linear map: $T_x \to T_x^*$ may be given by a covariant tensor of order two.

6 (5.1) Let U be an open set in E^n. Prove that the contravariant vector fields $\dfrac{\partial}{\partial x_1}, \ldots, \dfrac{\partial}{\partial x_n}$ are linearly independent over $\mathscr{D}(U)$.

7 (5.1) A curve $\gamma = (x_1(t), \ldots, x_n(t))$ in $U \subset E^n$ is called an *integral manifold* for a vector field $L = \{L_x\}$ on U if the tangent vector to γ at the point $x = \gamma(t)$ is the vector L_x. Given $L = \left\{x\dfrac{\partial}{\partial x} + y\dfrac{\partial}{\partial y}\right\}$ on E^2 find a family of integrable manifolds for L.

8 (7.1) Prove properties 7.2.

9 (7.1) Prove that for $L, M, N \in \mathscr{V}(U)$ the bracket [.,.] satisfies the properties:

$$[L, M] = -[M, L]$$

and

$$[L, [M, N]] + [M, [N, L]] + [N, [L, M]] = 0.$$

CHAPTER 3

Real manifolds

1.1 Differentiable varieties

ALTHOUGH MANIFOLDS are described as abstract geometrical objects, the idea of a manifold derives from rather concrete examples, namely non-singular varieties of E^n.

1.2 DEFINITION *A non-empty set $V \subset E^n$ is called a p-dimensional differentiable variety of E^n if to every point $x \in V$ there exist a neighborhood U of x in E^n and $m = n - p$ differentiable functions $\Phi_1, \dots, \Phi_m$ on U, such that:*

i) *the matrix* $\left(\dfrac{\partial \phi_i}{\partial x_j}\right)$, $i = 1, \dots, m; j = 1, \dots, n$, *has rank m at every point* $x \in U$,

ii) $V \cap U = \{x \in U \mid \Phi_1(x) = \cdots = \Phi_m(x) = 0\}$.

We give V the topology which is induced on it by the topology of E^n.

Let Ω be an open subset of E^n and let $f_1, \dots, f_m$ be m functions which are differentiable on Ω. Then the set:

$$S = \left\{x \in \Omega \mid f_1(x) = \cdots = f_m(x) = 0 \quad \text{and rank} \quad \left(\frac{\partial f_i}{\partial x_j}\right)_x = m\right\},$$

if not empty, is a p-dimensional differentiable variety of E^n. In fact, we can choose the same functions $f_1, \dots, f_m$ at every point $x \in S$ and observe that the set

$$U = \left\{x \in \Omega \mid \text{rank} \left(\frac{\partial f_i}{\partial x_j}\right)_x = m\right\}$$

is open. Then every $x \in S$ has a neighborhood U such that rank $\left(\dfrac{\partial f_i}{\partial x_j}\right)_x = m$ for all $x \in U$ and

$$S \cap U = \{x \in U \mid f_1(x) = \cdots = f_m(x) = 0\}.$$

It now follows by the implicit function theorem that, if Ω is an open subset of E^n and $f_1, \dots, f_r$ are r differentiable functions on Ω, then the set

$$V = \{x \in \Omega \mid f_1(x) = \cdots = f_r(x) = 0\}$$

is a p-dimensional differentiable variety of E^n if every point $x \in V$ has a neighborhood in E^n where the matrix $\left(\frac{\partial f_i}{\partial x_j}\right)$, $i = 1, \ldots, r; j = 1, \ldots, n$, has constant rank $m < n$ and m is the same at every $x \in V$.

By the local parametrization theorem, which is an easy consequence of the implicit function theorem, a p-dimensional differentiable variety V of E^n has the two following properties:

a) to every point $x \in V$ there exist an open subset $\Omega \subset E^p$ and an injective map $\Phi: \Omega \to E^n$ such that $\Phi(\Omega)$ is an open neighborhood of x in V,

b) there exist an open subset U of E^n, $U \supset \Phi(\Omega)$, and a differentiable map $\psi: U \to E^p$ such that $\psi \circ \Phi$ is the identity on Ω.

There are, of course, several choices for the set Ω and the map Φ. Every pair $(\Phi(\Omega), \Phi)$ is called a *coordinate neighborhood* of V. The map Φ and its inverse, the restriction of ψ to $\Phi(\Omega)$, determine a homeomorphism of Ω onto $\Phi(\Omega)$.

In the next section, properties (a), (b) will be generalized to topological spaces, not necessarily subsets of some Euclidean space. We shall however require these topological spaces to have a countable topology.

2.1 Differentiable manifolds

In chapter 2 the Euclidean space E^n had the role of providing, at every point $x \in E^n$, C^∞ local coordinates $x_1, \ldots, x_n$ on a neighborhood U of x and the ring $\mathscr{D}(U)$ of C^∞ functions on U. That was in fact sufficient to define at every $x \in E^n$ the tangent space T_x and from it all other related geometrical objects. It is important to remember that all C^∞ systems of coordinates we considered on an open subset U of E^n, namely all diffeomorphisms of U onto itself, are characterized by the condition of transforming $\mathscr{D}(U)$ onto itself.

Let M be a set. Let $\mathscr{D}$ be a ring of real valued functions defined on M which separate points of M; that is, if $m, m' \in M$, $m \neq m'$, there exists $f \in \mathscr{D}$ such that $f(m) \neq f(m')$. Give M the weakest topology such that all functions of $\mathscr{D}$ are continuous. For every $\varepsilon > 0$ and finite set $f_1, \ldots, f_k$ of functions of $\mathscr{D}$ the set

$$U_m(f_1, \ldots, f_k, \varepsilon) = \{p \in M \mid |f_i(m) - f_i(p)| < \varepsilon, i = 1, \ldots, k\}$$

is a neighborhood of m in M. Letting ε and the set $f_1, \ldots, f_k$ vary, we get a fundamental system of neighborhoods of $m \in M$.

A real valued function f, defined in M, is said to be *locally* in $\mathscr{D}$ if every point $m \in M$, where f is defined, has an open neighborhood U such that f is defined on U and, on U, f coincides with an element of $\mathscr{D}$.

2.2 Definition *The pair $(M, \mathscr{D})$ is called a differentiable manifold if the two following conditions are satisfied:*

i) *if a function f, defined on the whole of M, is locally in $\mathscr{D}$, then $f \in \mathscr{D}$,*

ii) *to every point $m \in M$ there is an open neighborhood U of m in M and a homeomorphism ϕ of an open subset $\Omega \subset E^n$ onto $U = \phi(\Omega)$ with the property that a function f defined on $\phi(\Omega)$ is locally in $\mathscr{D}$ if and only if $f \circ \phi \in C^\infty$ (Ω).*

Functions of $\mathscr{D}$ are called differentiable functions on M or $C^\infty(M)$. They are also said to define a C^∞ structure on M.

If $(M, \mathscr{D})$ is a differentiable manifold, we shall simply write M for $(M, \mathscr{D})$ when there is no doubt about the differentiable structure on M.

The integer n which appears in definition 2.2 is called the (topological) dimension of the differentiable manifold M.

2.3 *Remark* In condition (ii) of definition 2.2 we can confine ourselves to the consideration of the functions $f \circ \Phi$ which are analytic on Ω. In this case $(M, \mathscr{D})$ is called a *real analytic manifold.*

Let M be a differentiable manifold and U an open subset of M. Let $u_1, \ldots, u_n$ be n real valued functions on U defining an injective map $u = (u_1, \ldots, u_n) : U \to E^n$. Assume that u^{-1} is a homeomorphism with the property expressed by condition (ii) of definition 2.2. Then U is called a *coordinate neighborhood* of M, and u is called a *system of local coordinates* at every point $m \in U$ or a *coordinate system* for U. The pair (U, u) is also called a *local chart of M*.

2.4 *Remark* It follows from condition (ii) of definition 2.2 that (i) is equivalent to:

i') a function f, defined on all M, is in $\mathscr{D}$ if and only if M can be covered by local charts (U, u) such that $f \circ u^{-1}$ is C^∞ $(u(U))$.

2.5 *Remark* If (U, u) is a local chart of M, then the functions u_i are locally in $\mathscr{D}$. That the function $u_i \circ u^{-1}$ is C^∞ $(u(U))$ is clearly shown by the following diagram

$$\begin{array}{ccc} U & \xleftarrow{u^{-1}} & u(U) \subset E^n \\ {\scriptstyle u_i}\downarrow & \swarrow {\scriptstyle \pi_i | u(U)} & \\ \mathbf{R} & & \end{array}$$

where $\pi_i \colon E^n \to \mathbf{R}$ is the ith coordinate projection.

2.6 *Remark* It follows from the definition that every point m of a differentiable manifold M has a system of local coordinates; that is, if $m \in M$, there exists a local chart (U, u) of M with $m \in U$. However, in general, it is not possible to cover M with a unique chart (M, u).

2.7 THEOREM *Let $(M, \mathscr{D})$ be a differentiable manifold. Then every point $m \in M$ has a local chart (V, u) such that the u_i are functions of $\mathscr{D}$.*

Proof Let U be a coordinate neighborhood of m in M, and let $u = (u_1, \ldots u_n)$ be a coordinate system for U. We can assume that $u(m) = (0, \ldots, 0) \in E^n$ and that $u(U)$ is the open ball $B(r) \subset E^n$ with center at the origin and radius r.

Let $x_1, \ldots, x_n$ be coordinates in E^n and $x = (x_1, \ldots, x_n)$. For every $i = 1, \ldots, n$, the function $v_i(x) = u_i \circ u^{-1}(x)$ is C^∞ $(u(U))$. Let $B(r')$ be the open ball with center at the origin and radius $r' < r$. Let $\hat{v}_i(x)$ be a $C^\infty(E^n)$ function such that

$$\hat{v}_i(x) = \begin{cases} v_i(x) & \text{if} \quad x \in B(r') \\ v_i(0) & \text{if} \quad x \in E^n \backslash B(r). \end{cases}$$

Define now on M the function

$$f_i(m) = \begin{cases} \hat{v}_i \circ u(m) & \text{if} \quad m \in U \\ v_i(0) & \text{if} \quad m \in M \backslash U. \end{cases}$$

It is easy to check that the functions f_i are locally in $\mathscr{D}$ and therefore in $\mathscr{D}$. On the other hand, putting $V = u^{-1}(B(r'))$ we have $f_i \mid V = u_i$. ○

3.1 Second definition of a differentiable manifold

The definition of a differentiable manifold is usually given by "overlapping neighborhoods" in the following way:

Let M be a Hausdorff topological space with countable topology. Let U be an open set of M.

A *coordinate system* for U is a homeomorphism u of U onto an open subset of E^n. As before, U is called a *coordinate neighborhood*, u is called a *system of local coordinates* at every $m \in U$, and (U, u) is called a *local chart* of M.

Let (U, u) and (V, v) be two local charts of M. Then u, v are said to be *C^∞-related* if $u \circ v^{-1}$ and $v \circ u^{-1}$ are differentiable maps wherever they are defined.

Let $\mathscr{C} = \{(U_\alpha, u_\alpha)\}$ be a set of local charts of M.

3.2 DEFINITION *An n-dimensional differentiable manifold is a pair $(M, \mathscr{C})$ such that:*

j) $M = \cup U_\alpha$; *that is, every point $m \in M$ is contained in some coordinate neighborhood,*

jj) *for every α, β the maps u_α, u_β are C^∞-related,*

jjj) $\mathscr{C}$ *is maximal with respect to* (j) *and* (jj).

If $(M, \mathscr{C})$ is a differentiable manifold, then $\mathscr{C}$ is said to be a *differentiable structure* on M. Every subset of $\mathscr{C}$ which satisfies conditions (j) and (jj) in definition 3.2 is called a *basis* for the differentiable structure $\mathscr{C}$.

Let f be a real valued function defined on an open subset $U \subset M$. Then f is said to be $C^\infty(U)$ if, for every $m \in U$ and for every chart (V, v) with $m \in V \subset U$, the function $f \circ v^{-1} \colon v(V) \to \mathbf{R}$ is $C^\infty(v(V))$. It suffices, by condition (jj), that to every $m \in U$ there is a chart (V, v) such that $f \circ v^{-1}$ is $C^\infty(v(V))$.

If $\mathscr{C}_1, \mathscr{C}_2$ are two bases for some differentiable structures on M, then $\mathscr{C}_1$ and $\mathscr{C}_2$ are said to be equivalent if they define the same set of C^∞ functions. It is easy to check that $\mathscr{C}_1, \mathscr{C}_2$ are equivalent if and only if they are bases for the same differentiable structure on M. Condition (jjj) is meant to avoid the introduction of an equivalence relation.

3.3 *Remark* If $\mathscr{C} = \{(U_\alpha, u_\alpha)\}$ is a set of local charts of M satisfying only condition (j) of definition 3.2, then $(M, \mathscr{C})$ is called an *n-dimensional topological manifold.* On the same topological manifold there may be different differentiable structures. As an example, $(\mathbf{R}, t)$ and $(\mathbf{R}, t^3)$ are two different 1-dimensional differentiable manifolds. There are also examples of topological manifolds which do not admit any differentiable structure [12].

If (U_α, u_α), (U_β, u_β) are two charts of an n-dimensional differentiable manifold and $U_\alpha \cap U_\beta = U \neq \emptyset$, then the functions $u_\beta^\alpha = u_\beta \circ u_\alpha^{-1} \colon u_\alpha(U) \to u_\beta(U)$ and $u_\alpha^\beta = u_\alpha \circ u_\beta^{-1} \colon u_\beta(U) \to u_\alpha(U)$ are called *coordinate transformations* on

$U = U_\alpha \cap U_\beta$. They are both C^∞ functions and, since $u_\beta^\alpha = (u_\alpha^\beta)^{-1}$, their Jacobian matrices have rank n.

3.4 THEOREM *Definitions 2.2 and 3.2 are equivalent.*

Proof Let $(M, \mathscr{D})$ be a differentiable manifold. Let $\mathscr{C}$ be the set of all systems of coordinates (U, u) in M. We shall show that $\mathscr{C}$ is a differentiable structure on M. Condition (j) is clearly satisfied. In order to show that condition (jj) is satisfied, we must show that for any two systems of coordinates (U, u) and (V, v), the functions u and v are C^∞-related. Now, by remark 2.5, for every $i = 1, \ldots, n$ the function u_i is locally in $\mathscr{D}$ and thus $u_i \mid U \cap V$ is locally in $\mathscr{D}$. It follows then, by condition (ii), that $u_i \circ v^{-1} \in C^\infty(v(U \cap V))$. Therefore, $u \circ v^{-1} \in C^\infty(v(U \cap V))$. In the same way, one proves that $v \circ u^{-1} \in C^\infty(u(U \cap V))$.

To show that (jjj) is satisfied, it is enough to show that, if $(\tilde{U}, \tilde{u})$ is an arbitrary system of coordinates in M and $\tilde{u}$ is C^∞-related to every u of any system of coordinates $(U, u) \in \mathscr{C}$, then $(\tilde{U}, \tilde{u}) \in \mathscr{C}$. To every point $m \in \tilde{U}$ there exists a system of coordinates $(V, v) \in \mathscr{C}$, with $V \subset \tilde{U}$, such that a function f defined on V is locally in $\mathscr{D}$ if and only if $f \circ v^{-1} \in C^\infty(v(V))$. But now, by hypothesis, $f \circ v^{-1} \in C^\infty(v(V))$ if and only if $f \circ \tilde{u}^{-1} \in C^\infty(\tilde{u}(\tilde{U}))$.

Conversely, let $(M, \mathscr{C})$ be a differentiable manifold. Let $\mathscr{D}$ be the family of $C^\infty(U)$ functions where U is an open set in M. Since open sets of M and domains of definition of functions of $\mathscr{D}$ are also open sets in the weakest topology which makes all functions of $\mathscr{D}$ continuous, it follows that $(M, \mathscr{D})$ is a differentiable manifold. ○

4.1 Orientability

Let M be a differentiable manifold. Let (U, u), (V, v) be two charts of the differentiable structure $\mathscr{C}$ of M.

Let $U \cap V \neq \emptyset$. If $u = (u_1, \ldots, u_n)$ and $v = (v_1, \ldots, v_n)$, then the u_i are differentiable functions of the v_j and, if we put:

$$\left.\frac{\partial u_i}{\partial v_j}\right|_m = \left.\frac{\partial(u_i \circ v^{-1})}{\partial x_j}\right|_{x = v(m)},$$

then $\det(\partial u_i/\partial v_j)_m \neq 0$ at every $m \in U \cap V$. We shall say that (U, u) and (V, v) *agree in orientation* if $\det(\partial u_i/\partial v_j)_m > 0$ at every $m \in U \cap V$. It is, of course, enough to check that $\det(\partial u_i/\partial v_j)_m$ is positive at only one point for every connected component of $U \cap V$.

4.2 Definition *A differentiable manifold M is called orientable if there exists on M a differentiable structure $\mathscr{C}$ such that there is a basis of $\mathscr{C}$ in which every two charts agree in orientation.*

Later we shall give other conditions equivalent to the orientability.

If M is an orientable differentiable manifold, the choice of a basis of $\mathscr{C}$ such that every pair of charts in the basis agree in orientation is called an *orientation* for M.

If a connected differentiable manifold is orientable, then there are two possible orientations. An orientable differentiable manifold M is oriented when one of the orientations is chosen. If M is not connected, then M is orientable if every component of M is orientable and M is oriented when every component is oriented.

5.1 Examples

In this section we shall give some examples of differentiable manifolds. Most of the proofs will be left to the reader as an exercise.

5.2 *The topological space* $\mathbf{R}^n$ It is a real analytic manifold which can be covered by the unique chart (U, u) where $U = \mathbf{R}^n$ and $u =$ identity. $\mathbf{R}^n$ is a noncompact manifold.

5.3 *The sphere* $S^n \subset E^{n+1}$ It is a differentiable variety of E^{n+1}. S^n is an n-dimensional compact real analytic manifold which can be covered by the two charts (U_1, u_1), (U_2, u_2) where

$$U_1 = \{(x_1, \ldots, x_{n+1}) \in S^n \mid x_1 < 1\}$$

$$U_2 = \{(x_1, \ldots, x_{n+1}) \in S^n \mid x_1 > -1\}$$

and u_1, u_2 are the stereographic projections from $(1, 0, \ldots, 0)$ and $(-1, 0, \ldots, 0)$ respectively.

If $n = 2$, let S^2 be the sphere of the equation $x_1^2 + x_2^2 + x_3^2 = 1$ in E^3. Let $\mathscr{D}$ be the ring of functions defined on S^2, which, on a neighborhood of every point of S^2, are restrictions of C^∞ functions on some open subset of E^3.

The topology in S^2 is that which is induced on it by the topology of E^3.

Let U_1, U_2 be the open subsets of S^2 obtained by removing from S^2 the points $P_1 = (1, 0, 0)$ and $P_2 = (-1, 0, 0)$ respectively. Denote by u_1, u_2 the coordinates on the Euclidean space E^2 of the equation $x_1 = 0$.

Let ϕ_i, $i = 1, 2$ be the stereographic projections of E^2 onto S^2 from the points P_i. Then ϕ_i is a homeomorphism represented by

$$\phi_i: \begin{cases} x_1 = (-1)^i \dfrac{1 - u_1^2 - u_2^2}{1 + u_1^2 + u_2^2} \\[2ex] x_2 = \dfrac{2u_1}{1 + u_1^2 + u_2^2} \\[2ex] x_3 = \dfrac{2u_2}{1 + u_1^2 + u_2^2}. \end{cases}$$

Putting $E_i = E^3 \backslash \{P_i\}$, we have $U_i \subset E_i$. Let $\psi_i: E_i \to E^2$ be the map

$$\psi_i: \begin{cases} u_1 = \dfrac{x_2}{1 + (-1)^{i-1} x_1} \\[2ex] u_2 = \dfrac{x_3}{1 + (-1)^{i-1} x_1}. \end{cases}$$

Then clearly

$$\psi_i \mid U_i \circ \phi_i = \text{identity}.$$

Since condition (i) in definition 2.2 is obviously satisfied, in order to show that $(S^2, \mathscr{D})$ is a differentiable manifold we must show that a function f is locally in $\mathscr{D}$ if and only if $f \circ \phi_i \in C^\infty(E^2)$. If f is locally in $\mathscr{D}$, then every point $x \in S^2$, where f is defined, is contained in an open set $V \subset S^2$ such that f is defined on V and f is the restriction to V of a function $\hat{f} \in C^\infty(\Omega)$ for some open subset $\Omega \subset E^3$. We can assume $V \subset U_1$. Then $f \circ \phi_1 = \hat{f} \circ \psi_1^{-1}$ on $\phi_1^{-1}(V) = \psi_1(V)$ and thus $f \circ \phi_1 \in C^\infty(\phi_1^{-1}(V))$.

Conversely, let f be defined on $V \subset S^2$ such that $f \circ \phi_1 \in C^\infty(\phi_1^{-1}(V))$. Let $\hat{f} = f \circ \phi_1 \circ \psi_1 \in C^\infty(E_1)$. Evidently f is the restriction of $\hat{f}$ to V.

5.4 *The real projective space* $\mathbf{P}^n(\mathbf{R})$ It is the quotient space of $\mathbf{R}^{n+1} - \{0\}$ with respect to the equivalence relation

$$(x_1, \ldots, x_{n+1}) = (\lambda x_1, \ldots, \lambda x_{n+1})$$

where $\lambda \neq 0$ is any real number. $\mathbf{P}^n(\mathbf{R})$ is an n-dimensional compact real analytic manifold which can be covered by the $n + 1$ charts (U_i, u_i) where $U_i = \{(x_1, \ldots, x_{n+1}) \in \mathbf{P}^n(\mathbf{R}) \mid x_i \neq 0\}$ and $u_i: U_i \to \mathbf{R}^n$ is the homeomorphism which to $(x_1, \ldots, x_{n+1}) \in U_i$ associates the point

$$\left(\frac{x_1}{x_i}, \ldots, \frac{x_{i-1}}{x_i}, \frac{x_{i+1}}{x_i}, \ldots, \frac{x_{n+1}}{x_i}\right) \in \mathbf{R}^n.$$

5.5 *Differentiable varieties of* E^n are differentiable manifolds which need not be real analytic.

5.6 *Non-singular affine algebraic varieties* An affine algebraic variety is a set of points $\{x \in E^n \mid P_1(x) = \cdots = P_r(x) = 0\}$ where P_i is a polynomial in $x = (x_1, \ldots, x_n)$. A non-singular affine algebraic variety can always be given the structure of a real analytic manifold.

5.7 *Non-singular projective algebraic varieties* A projective algebraic variety is a set of points

$$\{x \in \mathbf{P}^n(\mathbf{R}) \mid P_1(x) = \cdots = P_r(x) = 0\}$$

where P_i is a homogeneous polynomial in $x = (x_1, \ldots, x_{n+1})$. A non-singular projective algebraic variety can always be given the structure of a compact real analytic manifold.

5.8 2-*dimensional torus* T^2 As a topological manifold, T^2 is the product of two 1-dimensional spheres. It is a non-singular algebraic variety of E^3 and can be given the corresponding structure of a real analytic manifold.

Not all the manifolds of the above examples are orientable. For instance, $\mathbf{R}^n$ and S^n are orientable, while $\mathbf{P}^n(\mathbf{R})$ is orientable if and only if n is odd.

Examples of differentiable manifolds arise by a process called *identification*. A topological space M can be made into a differentiable manifold if there is a homeomorphism of M onto a differentiable variety V of E^n. The given homeomorphism can be used to transfer the differentiable structure from V to M. The n-dimensional sphere S^n can be easily produced by this process.

Other examples of differentiable manifolds are given by *products*. Let $(M, \mathscr{C})$ and $(M', \mathscr{C}')$ be two differentiable manifolds. We give the topological product $M \times M'$ the structure of a differentiable manifold. Let $\pi\colon M \times M' \to M$ and $\pi'\colon M \times M' \to M'$ be the natural projections. Let (U, u) and (U', u') be two charts of M and M' respectively. Define $u \times u' = (u \circ \pi, u' \circ \pi')$. Then the set $\mathscr{C}_0 = \{(U \times U', u \times u')\}$ is a basis of a differentiable structure on $M \times M'$. In particular, an n-dimensional torus T^n can be made into an n-dimensional differentiable manifold, as in example 5.8 above for $n = 2$.

More examples of differentiable manifolds may be produced by the following construction. Let M be a differentiable manifold. A group Γ is called a *group of automorphisms* of M if the elements $\gamma \in \Gamma$ are injective maps $\gamma\colon M \to M$ such that the differentiable structures of M and γM coincide. Γ is said to be *properly discontinuous* if every point $m \in M$ has a neighborhood U which does not intersect γU if $\gamma \neq$ identity of Γ. The group Γ acts on M with no fixed points.

Let Γ be a properly discontinuous group acting on the differentiable manifold M with no fixed points. It is easy to see that M/Γ is a differentiable manifold of the same dimension as M. As an example, the 2-dimensional torus T^2 can be described in the following way: Let $x = (x_1, x_2)$ and $y = (y_1, y_2)$ be two points in E^2 such that, under the identification of E^2 and V^2, x and y form a basis of V^2. Let Γ be the subgroup of V^2 consisting of all linear combinations $ax + by$ with $a, b \in \mathbf{Z}$. Then Γ acts on E^2 as a properly discontinuous group of automorphisms and it is easy to see that E^2/Γ is the torus T^2.

6.1 Germs of differentiable functions

Let M be a differentiable manifold and let m be a point of M. We say that a function f is differentiable *near m* if f is differentiable on an open neighborhood of m. We shall now introduce an equivalence relation among the functions which are differentiable near m. Two functions f and g will be considered equivalent if they have the same value on a neighborhood of m. Each equivalence class will be called a *germ* of differentiable function at m. The set of all germs at $m \in M$ will be denoted by $\mathscr{F}_m$. The germ of f will be denoted by $\mathscr{f}$. We remark that the germ of a function at m depends on the behavior of the function in an open neighborhood of m and not only on the value of the function at m.

All functions of the same germ have the same value at m. This value is called the value of the germ at m.

Let f_1 and f_2 be two differentiable functions near m and let $\mathscr{f}_1$ and $\mathscr{f}_2$ be their germs. On some neighborhood of m the sum $f_1 + f_2$ is defined and it is a differentiable function near m. Then we can define $\mathscr{f}_1 + \mathscr{f}_2$ to be the germ of the function $f_1 + f_2$. Similarly, if f is differentiable near m and $\mathscr{f}$ is its germ, for every constant c, cf is differentiable near m and we can define $c\mathscr{f}$. These operations are well defined and they make $\mathscr{F}_m$ into a vector space which is easily seen to be of infinite dimension.

If $\mathscr{f}, \mathscr{g}$ are two germs of $\mathscr{F}_m$, the product of the two germs can be defined as the germ of the product of two representatives f, g of $\mathscr{f}, \mathscr{g}$ since it is easy to show that it does not depend on the choice of f and g.

Hence $\mathscr{F}_m$ is made into a commutative algebra.

Of course, germs of differentiable functions could have been already introduced in E^n.

6.2 THEOREM *Let M be a differentiable manifold. Let m, m' be two points of M. There is an algebra isomorphism between $\mathscr{F}_m$ and $\mathscr{F}_{m'}$.*

Proof Let (U, u), (U', u') be two local charts of M containing m and m' respectively. We can assume that $u(m) = u'(m')$. Let $\not{f} \in \mathscr{F}_m$ and let f be a representative of $\not{f}$. Then f is a differentiable function on some open set contained in U. Consider the function $g = f \circ u^{-1} \circ u'$ which is defined on a neighborhood of m'. Since $g \circ u'^{-1} = f \circ u^{-1}$ is a differentiable function on a neighborhood of $u'(m')$, then g is differentiable near m'. Let $\not{g}$ be the germ of g in $\mathscr{F}_{m'}$ and let $\alpha: \mathscr{F}_m \to \mathscr{F}_{m'}$ be the map $\alpha(\not{f}) = \not{g}$. We must show that α is well defined. Let f_1 be another representative of $\not{f}$. Then $f - f_1 = 0$ on a neighborhood of m. If $g_1 = f_1 \circ u^{-1} \circ u'$, then $g - g_1 = (f - f_1) \circ u^{-1} \circ u' = 0$ on a neighborhood of m' and thus g and g_1 belong to the same germ $\not{g}$.

We leave it to the reader to verify that α preserves operations. ○

Let M, N be two differentiable manifolds. A map $\phi: M \to N$ is called a *differentiable map* or a C^∞ map if, for every two charts (U, u) and (V, v) of M and N respectively, the function $v \circ \phi \circ u^{-1} \in C^\infty(u(U))$.

It turns out that for $\phi: M \to N$ to be differentiable it is enough that to every point $m \in M$ and $\phi(m) \in N$ there are local charts (U, u) and (V, v) containing m and $\phi(m)$ respectively, such that $v \circ \phi \circ u^{-1} \in C^\infty(u(U))$.

It follows that $\phi: M \to N$ is a differentiable map if and only if the local coordinates at $\phi(m) \in N$ are differentiable functions of the local coordinates at $m \in M$.

Let $\phi: M \to N$ be a differentiable map, let $m \in M$, and let (U, u) be a local chart of M containing m. If f is a differentiable function on a neighborhood of $\phi(m)$, when no confusion is possible we shall write $\left.\dfrac{\partial f}{\partial u_i}\right|_m$ for $\left.\dfrac{\partial (f \circ \phi)}{\partial u_i}\right|_m$.

6.3 DEFINITION *An injective and surjective map $\phi: M \to N$ is called a diffeomorphism if both ϕ and ϕ^{-1} are differentiable maps.*

Following the line at the beginning of this section, germs of differentiable maps and of diffeomorphisms can be easily defined.

7.1 Topological structure of manifolds

The structure of a topological manifold M is usually given by assigning an open covering $\mathscr{U} = \{U_\alpha\}_{\alpha\in A}$ of M and a system of local coordinates u_α on every open set U_α. If M is a differentiable manifold, then on every non-empty intersection $U_\alpha \cap U_\beta$ the coordinates u_α must be differentiable functions of the coordinates u_β and vice versa.

Let M be a topological space. An open covering $\mathscr{U} = \{U_\alpha\}_{\alpha\in A}$ of M is called *locally finite* if every $m \in M$ has a neighborhood which meets only a finite number of U_α.

Let $\mathscr{U} = \{U_\alpha\}_{\alpha\in A}$ be an open covering of the topological space M. Let $\mathscr{V} = \{V_\beta\}_{\beta\in B}$ be a second open covering of M. $\mathscr{V}$ is said to be a *refinement* of $\mathscr{U}$ if every V_β is contained in some U_α.

A topological space M is called *paracompact* if every open covering of M has a locally finite refinement.

7.2 THEOREM *Every locally compact Hansdorff topological space M with countable topology is paracompact.*

Proof Since M has countable topology, M has a countable basis $\{U_i\}$ of open sets. Since M is locally compact, we may assume that every $\bar{U}_i$* is compact.

Define $K_1 = \bar{U}_1$. Assume that the compact set K_{m-1} is defined and let $j_m > j_{m-1}$ be the least integer such that

$$K_{m-1} \subset \bigcup_{j=1}^{j_m} U_j.$$

Then define

$$K_m = \bigcup_{j=1}^{j_m} \bar{U}_j.$$

We have that $K_{m-1} \subset \mathring{K}_m$† and that every compact subset of M is contained in K_m if m is big enough.

Define now

$$V_i = U_i \quad \text{if} \quad i \leq j_2,$$

and

$$V_i = U_i \cap CK_{m-1}\text{††} \quad \text{if } j_m < i \leq j_{m+1} \; (m > 1).$$

In order to show that $\mathscr{V} = \{V_i\}$ is a covering of M, it is enough to show that

$$\bigcup_{i=1}^{j_{m+1}} V_i = \bigcup_{i=1}^{j_{m+1}} U_i.$$

---

* $\bar{U}_i$ denotes the closure of V_i.

† $\mathring{K}_m$ denotes the interior of K_m.

†† CK_{m-1} denotes the complement of K_{m-1}.

The above relation may be proved by induction. It is true for $m = 1$ since $V_i = U_i$ if $i \leq j_2$. If we now assume that

$$\bigcup_{i=1}^{j_m} V_i = \bigcup_{i=1}^{j_m} U_i = A$$

for some $m > 1$, since $K_{m-1} \subset A$, it follows at once that

$$V_i \cup A = U_i \cup A \quad \text{for} \quad j_m < i \leq j_{m+1}.$$

Moreover, the covering $\mathscr{V}$ is locally finite since every compact set is contained in K_m for m big enough and V_i does not meet K_m if $i > j_{m+1}$. ○

If M is a topological manifold, then M is a locally compact topological space. We also assume that M has countable topology. Therefore, by the above theorem we can always replace the covering $\mathscr{U}$ of M with a locally finite covering. Thus, from now on we shall assume that every covering we consider is locally finite. This assumption has another important consequence as stated in the following theorem.

7.3 THEOREM *Let M be a topological manifold. To every locally finite open covering $\mathscr{U} = \{U_\alpha\}_{\alpha \in A}$ of M, there exists a locally finite open covering $\mathscr{V} = \{V_\alpha\}_{\alpha \in A}$ with the same index set, such that for every $\alpha \in A$, $\bar{V}_\alpha \subset U_\alpha$.*

Proof For every $m \in M$ we can find an open set W containing m such that $\bar{W} \subset U_\alpha$ for some $\alpha \in A$. The new covering $\mathscr{W} = \{W_\beta\}$ of M may not be locally finite, but, since M is paracompact, we can find a refinement $\mathscr{W}'$ of $\mathscr{W}$ such that $\mathscr{W}'$ is locally finite and $\bar{W}'_\beta \subset U_\alpha$.

Now define

$$V_\alpha = \bigcup_{W'_\beta \subset U_\alpha} W'_\beta.$$

Then we have $\mathscr{V} = \{V_\alpha\}_{\alpha \in A}$ locally finite and $\bar{V}_\alpha \subset U_\alpha$.

If f is a function defined on an open set U, the *support* of f, denoted by supp f, is the smallest closed set such that $f(x) = 0$ if $x \notin \operatorname{supp} f$.

7.4 LEMMA *Let K be a compact subset of E^n and let U be an open set containing K. There exists a C^∞ function ϕ, with compact support contained in U, such that $0 \leq \phi(x) \leq 1$ for all x and $\phi(x) = 1$ for $x \in K$.*

Proof Assume first that K is the closed ball $|x| \leq r$ and U is the open ball $|x| < R$ with $0 < r < R$. The function

$$g(t) = \begin{cases} e^{-(t-r)^{-1}} e^{-(t-R)^{-1}} & \text{if} \quad r < t < R \\ 0 & \text{otherwise} \end{cases}$$

is a differentiable function. Then the function

$$f(x) = \frac{\int_{|x|}^{R} g(t)\,dt}{\int_{r}^{R} g(t)\,dt}$$

is also differentiable and $f(x) = 1$ if $|x| \leq r$, and $f(x) = 0$ if $|x| \geq R$.

In general, let $K_i \subset U_i$ be concentric solid balls such that $K \subset \bigcup_i K_i$ and $U_i \subset U$. Let $f_i(x)$ be the function which satisfies the conditions of the lemma for K_i, U_i. By the compactness of K, only a finite number of pairs K_i, V_i are needed. Thus the function

$$\phi(x) = 1 - \prod_i (1 - f_i(x))$$

is well defined and satisfies the same conditions for K and U. ○

7.5 Definition *Let $\mathscr{U} = \{U_\alpha\}$ be a locally finite open covering of the differentiable manifold M. A set $\{\phi_\alpha\}$ of C^∞ functions on M is called a C^∞ partition of unity subordinate to the covering $\mathscr{U}$ if:*

i) $\phi_\alpha \geq 0$ *for all* α,

ii) supp $\phi_\alpha \subset U_\alpha$,

iii) $\sum_\alpha \phi_\alpha(m) = 1$ *for every* $m \in M$.

7.6 Lemma *Let M be a differentiable manifold and let $\mathscr{U}$ be a locally finite open covering of M. There exists a C^∞ partition of unity subordinate to the covering $\mathscr{U}$.*

Proof Let $\mathscr{V}$ be the refinement of $\mathscr{U}$ as in theorem 7.3. By lemma 7.4 there exists a C^∞ function f_α, such that

$$f_\alpha(m) = \begin{cases} 1 & \text{if } m \in \bar{V}_\alpha \\ 0 & \text{if } m \notin U_\alpha. \end{cases}$$

For every $m \in M$ we can define

$$f(m) = \sum_{\alpha=1}^{\infty} f_\alpha(m)$$

since the sum has only a finite number of terms different from zero.

We can now define

$$\phi_\alpha(m) = \frac{f_\alpha(m)}{f(m)}$$

and then have $\phi_\alpha(m) \geq 0$ for all α, supp $\phi_\alpha \subset U_\alpha$ and $\sum_\alpha \phi_\alpha(m) = 1$ for all $m \in M$. ○

7.7 *Remark* If M is a topological manifold, then to every locally finite open covering $\mathscr{U}$ of M there exists a partition of unity subordinated to $\mathscr{U}$ with the ϕ_α continuous functions on M.

If M is a real analytic manifold, there are, of course, C^∞ partitions of unity subordinated to any locally finite open covering of M, but, in general, it is not possible to ask that the ϕ_α be real analytic functions on M.

7.8 THEOREM *Let M be a differentiable manifold and let $m \in M$. Every germ in $\mathscr{F}_m$ is represented by a function which is differentiable on all of M.*

Proof Let f be a function defined on M and differentiable near m. Let (U, u) be a chart of M containing m and let $0, \ldots, 0$ be the coordinates of m. Let K be a compact set contained in U. Let g be a C^∞ function on U such that supp $g \subset U$, $g(m') = 1$ for $m' \in K$, and $0 \leq g \leq 1$ on U, as in lemma 7.4.

Define the function $\hat{f}$ on M by

$$\hat{f}(m') = \begin{cases} f(u^{-1}[g(m')\, u_1(m'), \ldots, g(m')\, u_n(m')]) & \text{if} \quad m' \in U \\ f(m) & \text{if} \quad m' \in M \backslash U. \end{cases}$$

Then $\hat{f}$ is differentiable on all M and, since $\hat{f} \mid K = f$, then f and $\hat{f}$ belong to the same germ of $\mathscr{F}_m$. ○

Let $\phi\colon M \to N$ be a differentiable map. From the previous theorem it follows that the germ of ϕ at every $m \in M$ can be represented by a map which is differentiable on all M.

8.1 Tangent space

Let M be a differentiable manifold of dimension n and let m be a point of M. Since m has a neighborhood which is homeomorphic to an open subset of the n-dimensional Euclidean space, we can define the tangent space to M at m as we did for a point of E^n in 2.3.1. In the case of a manifold M, the tangent space gives locally a linear approximation of the topological space M. As we have already remarked, tangent vectors depend on the behavior of differentiable functions near m and not only on the values of the functions at m. Thus, we do not have to distinguish among functions belonging to the same germ of $\mathcal{F}_m$. If f is a representative of the germ $f \in \mathcal{F}_m$, we shall simply write $f \in \mathcal{F}_m$.

8.2 DEFINITION *A tangent vector at $m \in M$ is a functional $t: \mathcal{F}_m \to \mathbf{R}$ such that:*

$$t(f + g) = t(f) + t(g),$$

$$t(fg) = f(m)\, t(g) + g(m)\, t(f),$$

$$t(f) = 0 \quad \textit{if} \quad f = \textit{constant},$$

for $f, g \in \mathcal{F}_m$.

The *tangent space* to M at m is the vector space $T_m(M)$ of all tangent vectors at $m \in M$. We shall write T_m for $T_m(M)$ when no confusion is possible. T_m is an n-dimensional vector space. If $u = (u_1, \ldots, u_n)$ is a coordinate system on a neighborhood of m, a basis of T_m is given by $\left\{\frac{\partial}{\partial u_i}\right\}_{i=1,\ldots,n}$. Then, for every $t \in T_m$ we have

$$t = \sum_{i=1}^{n} t^i \frac{\partial}{\partial u_i},$$

with $t^i = t(u_i)$, $i = 1, \ldots, n$, and for every $f \in \mathcal{F}_m$,

$$\frac{\partial f}{\partial u_i}(m) = \frac{\partial (f \circ u^{-1})}{\partial x_i}(u(m)).$$

Tangent vectors have the following geometrical characterization. A differentiable curve in M is any map $\gamma: I \to M$ of the closed interval $I = [-1, 1]$ into M which can be extended to a differentiable map of an open interval.

Let γ be a differentiable curve in M such that $\gamma(0) = m$. Define a linear functional $\gamma_*: \mathcal{F}_m \to \mathbf{R}$ by:

$$\gamma_*(f) = \left.\frac{d(f \circ \gamma)}{d\tau}\right|_{\tau=0}$$

for $f \in \mathcal{F}_m$, $\tau \in I$.

γ_* is clearly a tangent vector at $\gamma(0) = m \in M$; that is, γ_* satisfies the conditions of definition 8.2. We shall show now that every tangent vector of T_m is of the form γ_* for some differentiable curve γ such that $\gamma(0) = m$. In fact, if $t \in T_m$ and $t = \sum_{i=1}^{n} t^i \frac{\partial}{\partial u_i}(m)$, then, clearly t is the vector tangent to the curve $\gamma: \tau \to u^{-1}[u_1(m) + \tau t^1, \ldots, u_n(m) + \tau t^n]$ at the point $\gamma(0)$.

We remark that the partial derivatives of a function f with respect to the local coordinates u_i at a point $m \in M$ depend on the choice of the local chart (U, u) containing m and therefore have no intrinsic meaning on M. The number $\gamma_*(f)$, which can be thought of as the directional derivative of f at m in the direction of γ, clearly does not depend on the choice of the chart (U, u). Let $\gamma(\tau)$ and $\gamma_1(\tau_1)$ be two differentiable curves in M, and assume that $\gamma(0) = \gamma_1(0) = m \in M$. Then, for $f \in \mathscr{F}_m$, we have $\gamma_*(f) = \gamma_{1*}(f)$ if and only if

$$\left.\frac{du_i}{d\tau}\right|_{\tau=0} = \left.\frac{du_i}{d\tau_1}\right|_{\tau_1=0}$$

for $i = 1, \ldots, n$. It follows that, for the consideration of tangent vectors at m, we have to identify two curves γ, γ_1 if they satisfy the above equalities, which clearly do not depend on the local chart (U, u).

Let γ_* be a tangent vector of T_m and let (U, u) be a chart containing $m = \gamma(0)$. The map

$$\gamma_* \to \left.\left(\frac{du_1}{d\tau}, \ldots, \frac{du_n}{d\tau}\right)\right|_{\tau=0}$$

is an injective map $T_m \to \mathbf{R}^n$. This map is not natural. In fact, if (V, v) is another chart with $m \in U \cap V$, we have:

$$\left.\left(\frac{du_i}{d\tau}\right)\right|_{\tau=0} = \left(\frac{\partial u_i}{\partial v_j}\right)_m \left.\left(\frac{dv_j}{d\tau}\right)\right|_{\tau=0}.$$

However, since the relation between

$$\left.\left(\frac{du_1}{d\tau}, \ldots, \frac{du_n}{d\tau}\right)\right|_{\tau=0} \quad \text{and} \quad \left.\left(\frac{dv_1}{d\tau}, \ldots, \frac{dv_n}{d\tau}\right)\right|_{\tau=0}$$

is linear homogeneous, the vector space structure of $\mathbf{R}^n$ may be transferred to T_m in a natural way, that is, independent of the local coordinates at m. Then, at every $m \in M$, a basis of T_m is the canonical basis B_0 of V^n.

9.1 Cotangent space

The *cotangent space* to M at m is the vector space $T_m^*(M)$, dual of the tangent space $T_m(M)$. We shall write T_m^* for $T_m^*(M)$ when no confusion is possible. T_m^* is an n-dimensional vector space; its elements are called *cotangent vectors* at $m \in M$. The basis of T_m^*, dual of the basis

$$\left\{\frac{\partial}{\partial u_i}\right\}_{i=1,\dots,n} \text{ of } T_m, \text{ is } \{du^i\}_{i=1,\dots,n}.$$

These bases are, of course, induced by the same system of local coordinates $u = (u_1, \dots, u_n)$ at m. Then, for every $t^* \in T_m^*$, we have:

$$t^* = \sum_{i=1}^{n} t_i^* du^i$$

with $t_i^* = t^*\left(\dfrac{\partial}{\partial u_i}\right)$.

Let (U, u) be a chart of M and let $m \in U$. For every $f \in \mathscr{F}_m$ we have the map $d: \mathscr{F}_m \to T_m^*$ defined by $\langle t, df\rangle = \langle f, t\rangle$ for $t \in T_m$. Moreover, if f, g are elements of $\mathscr{F}_m$ such that $df = dg$, then $\langle t, df\rangle = \langle t, dg\rangle$ for every $t \in T_m$.

In particular, for $t = \dfrac{\partial}{\partial u_i}$, $f = u_j$, we have:

$$\left\langle \frac{\partial}{\partial u_i}, \; du^j \right\rangle = \delta_i^j$$

and, for every $f \in \mathscr{F}_m$,

$$df = \sum_{i=1}^{n} \frac{\partial f}{\partial u_i} du^i.$$

The *differential* df at $m \in M$ of a differentiable function f near m has an intrinsic meaning on M; that is, it does not depend on the choice of the local coordinates at m. If f, g are two differentiable functions near $m \in M$, it is clear that $df = dg$ if $\langle t, df\rangle = \langle t, dg\rangle$ for every $t \in T_m$. It follows that f and g have the same differential at m if and only if, whenever we fix a local chart (U, u) containing m, we have $\dfrac{\partial f}{\partial u_i}(m) = \dfrac{\partial g}{\partial u_i}(m)$ for $i = 1, \dots, n$. Thus, in the chosen chart, the n-tuple $\left(\dfrac{\partial f}{\partial u_1}(m), \dots, \dfrac{\partial f}{\partial u_n}(m)\right)$ determines df at m.

If (V, v) is another chart of M with $m \in U \cap V$, we have:

$$\left(\frac{\partial f}{\partial u_i}\right)_m = {}^t\left(\frac{\partial v_j}{\partial u_i}\right)_m \left(\frac{\partial f}{\partial v_j}\right)_m$$

which shows, as in the previous section, that the vector space structure of T_m^* is natural.

10.1 Vector fields on a manifold

Let U be an open set of the differentiable manifold M. A *contravariant vector field* $X = \{X_m\}$ *on* U is a function which to every point $m \in U$ assigns a vector $X_m \in T_m$.

Let $u_1, \ldots, u_n$ be local coordinates at m; then, in a neighborhood of m we have:

$$X = \sum_{i=1}^{n} X^i (u_1, \ldots, u_n) \frac{\partial}{\partial u_i}.$$

If X is a contravariant vector field on U, as in section 2.5.1, we define, for every $f \in \mathscr{D}(U)$, the function Xf by $Xf(m) = X_m(f)$. The vector field X is said to be C^∞ if, for every $f \in \mathscr{D}(U)$, the function $Xf \in \mathscr{D}(U)$. It follows that X is C^∞ if and only if the n functions X^i are C^∞.

10.2 THEOREM *For every C^∞ contravariant vector field X on U the map $f \to Xf$ is a derivation on $\mathscr{D}(U)$. Conversely, to every map D: $\mathscr{D}(U) \to \mathscr{D}(U)$ such that:*

$$D(f+g) = D(f) + D(g),$$

$$D(fg) = fD(g) + gD(f),$$

$$D(f) = 0 \quad \textit{if} \quad f = \textit{constant},$$

there exists a unique C^∞ contravariant vector field X on U such that $D(f) = Xf$ for every $f \in \mathscr{D}(U)$.

The proof is the same as in section 2.5.3. ○

The set of all contravariant vector fields on U, $\mathscr{V}(U)$ is easily made into a module over $\mathscr{D}(U)$ and into a Lie algebra with the product

$$[X, Y]f = X(Yf) - Y(Xf)$$

for $X, Y \in \mathscr{V}(U)$ and $f \in \mathscr{D}(U)$.

Moreover, the multiplication fX defined by

$$(fX)_m = f(m)\, X_m$$

satisfies the following property:

$$[fX, gY] = fg\,[X, Y] + f(Xg)\, Y - g\,(Yf)\, X$$

for $X, Y \in \mathscr{V}(U)$ and $f, g \in \mathscr{D}(U)$.

A *covariant vector field* $X^* = \{X_m^*\}$ on an open set $U \subset M$ is a function which to every $m \in U$ assigns a vector $X_m^* \in T_m^*$.

If $u_1, \ldots, u_n$ are coordinates at m, then in a neighborhood of m we have:

$$X^* = \sum_{i=1}^{n} X_i^* (u_1, \ldots, u_n) \, du^i .$$

A covariant vector field X^* on U is said to be C^∞ if, for every contravariant vector field X on U, the function

$$\langle X, X^* \rangle (m) = \langle X_m, X_m^* \rangle$$

is C^∞. Hence, X^* is C^∞ if and only if the n functions X_i^* are C^∞.

If $f \in \mathscr{D}(U)$, the *differential* of f is the covariant vector field

$$df = \{(df)_m\},$$

where

$$(df)_m (t) = \langle f, t \rangle$$

for $t \in T_m$. Then for every $f \in \mathscr{D}(U)$ one has:

$$df = \sum_{i=1}^{n} \frac{\partial f}{\partial u_i} \, du^i .$$

In particular, the vector field du^i is the differential of the projection $\pi_i(m) = u_i(m)$.

10.3 *Remark* The definition of a C^∞ vector field on an open set $U \subset M$ does not depend on the local charts. In particular, we may take $U = M$. The notion of a C^∞ vector field on M has an intrinsic meaning.

10.4 *Remark* On every manifold M there exists some C^∞ contravariant vector field; for example, the 0 vector field. In order that a manifold M has some C^∞ non-vanishing contravariant vector field, M must satisfy some topological conditions. We shall not consider this problem. As an example, if M is compact, a necessary and sufficient condition for the existence of a non-vanishing C^∞ contravariant vector field on M is the vanishing of the Euler characteristic of M (Hopf theorem).

Definitions of tensor fields on a manifold can be given following the lines of section 2.7.1. We leave it to the reader to formulate these definitions and study the properties of tensor fields.

11.1 Differential forms

Let m be a point of the n-dimensional differentiable manifold M. We construct the Grassmann algebra of the vector space T_m^* and put $A_m^p = \wedge^p T_m^*$. Then we have:

$$\wedge T_m^* = A_m = A_m^0 \oplus A_m^1 \oplus \cdots \oplus A_m^n,$$

where $A_m^0 = \mathbf{R}$ and $A_m^1 = T_m^*$.

If (U, u) is a local chart of M with $m \in U$, the vector space A_m^p has the natural basis

$$\{du^{i_1} \wedge \cdots \wedge du^{i_p}\}_{1 \leq i_1 < \cdots < i_p \leq n}.$$

With respect to this basis, an element $\omega^p(m) \in A_m^p$ has a unique representation as

$$\omega^p(m) = \sum_{1 \leq i_1, < \cdots < i_p \leq n} \omega_{i_1 \ldots i_p} du^{i_1} \wedge \cdots \wedge du^{i_p}$$

with $\omega_{i_1 \cdots i_p} \in \mathbf{R}$.

Let W be an open subset of M. Assume we have a function which to every point $m \in W$ associates an element $\omega^p(m) \in A_m^p$ so that in every local chart the coefficients $\omega_{i_1 \cdots i_p}$ are differentiable functions of the coordinates of m. Then we say that the $\omega^p(m)$ determine a C^∞ *differential form* ω^p on W. The differential form ω^p is said to be *homogeneous of degree p*. By differential form we shall always mean C^∞ differential form.

More generally, we may consider a function which to every point $m \in W$ associates an element $\omega(m) \in A_m$, with the same condition for the coefficients. Then the $\omega(m)$ determine a differential form ω on W which is the sum of homogeneous differential forms. In what follows we shall consider only homogeneous differential forms.

Clearly, the consideration of differential forms on an open set W of M is the same as the consideration of C^∞ antisymmetric covariant tensor fields on W.

In a fixed chart (U, u) a differential form ω^p has the expression

11.2 $$\omega^p = \sum_{1 \leq i_1 < \cdots < i_p \leq n} \omega_{i_1 \ldots i_p} du^{i_1} \wedge \cdots \wedge du^{i_p}$$

where $\omega_{i_1 \cdots i_p}$ are C^∞ functions on U. If (V, v) is another chart, then we have

$$\omega^p = \sum_{1 \leq i_1 < \cdots < i_p \leq n} \omega'_{i_1 \ldots i_p} dv^{i_1} \wedge \cdots \wedge dv^{i_p}$$

where $\omega'_{i_1 \cdots i_p}$ are C^∞ functions on V. Assume $U \cap V \neq \phi$. From the transformation law of the components of covariant tensors and that of cotangent vectors, we have that the coefficients of ω^p transform with the law:

$$\omega'_{i_1 \cdots i_p} = \sum_{j_1 \cdots j_p} \omega_{j_1 \cdots j_p} \frac{\partial u_{j_1}}{\partial v_{i_1}} \cdots \frac{\partial u_{j_p}}{\partial v_{i_p}}.$$

In this expression the sum is taken over all non-ordered p-tuples $j_1 \ldots j_p$ and $\omega_{j_1 \cdots j_p} = \pm \omega_{k_1 \cdots k_p}$. Here the sign is $+$ if $j_1 \ldots j_p$ is an even permutation of $k_1 \ldots k_p$ and otherwise it is $-$.

A differential form ω^p vanishes at a point $m \in M$ if every coefficient $\omega_{i_1 \cdots i_p}$ vanishes at m. Two differential forms are equal if and only if their difference vanishes everywhere. The support of a differential form ω^p, denoted by supp ω^p, is the smallest closed set such that $\omega^p(m) = 0$ if $m \notin \operatorname{supp} \omega^p$.

Let V be an open subset of M. We shall denote by $A^p(V)$ the vector space of differential forms of degree p defined on V. We shall define a differential operator:

$$d: A^p(V) \to A^{p+1}(V).$$

This operator d is called the *exterior differential* and it is defined locally as follows: In a local chart (U, u) a differential form $\omega^p \in A^p(V)$ has the expression 11.2. The differential of ω^p is the form of degree $p + 1$,

$$d\omega^p = \sum_{1 \leq i_1 < \cdots < i_p \leq n} (d\omega_{i_1 \cdots i_p}) \wedge du^{i_1} \wedge \cdots \wedge du^{i_p}.$$

If ω^0 is a differential form of degree zero, $d\omega^0$ is the usual differential of the function ω^0, that is,

$$d\omega^0 = \sum_{i=1}^{n} \frac{\partial \omega^0}{\partial u_i} du^i.$$

We remark that d is defined locally and not punctually, that is, it does not have any meaning on A^p_m for fixed m.

The exterior differential satisfies the following properties:

i) for any two differential forms ω_1, ω_2 defined on V, we have

$$d(\omega_1 + \omega_2) = d\omega_1 + d\omega_2,$$

which shows that d preserves the vector space structure;

ii) If ω^p, ω^q are differential forms on V, of degree p, q respectively, then

$$d(\omega^p \wedge \omega^q) = d\omega^p \wedge \omega^q + (-1)^p \omega^p \wedge d\omega^q,$$

which shows that d does not preserve the algebra structure;

iii) For every differential form ω, we have

$$d^2\omega = 0.$$

Properties (i), (ii), (iii) show that the definition of $d\omega^p$, which *a priori* seems to depend on the local coordinates, is actually independent of them. Thus, if ω^p is a differential form defined on any open set $W \subset M$, $d\omega^p$ is a well defined differential form on W. We can, of course, take $W = M$.

11.3 DEFINITION *A differential form ω is called closed if $d\omega = 0$. A differential form ω^p is called exact if there exists a differential form ω^{p-1} such that $d\omega^{p-1} = \omega^p$.*

It follows from property (iii) that every exact differential form is closed.

12.1 Action of maps

Let M^r, N^s be two differentiable manifolds of dimension r, s respectively and let $\phi: M^r \to N^s$ be a differentiable map. The map ϕ induces a map from the tangent space to M^r at a point $m \in M^r$ into the tangent space to N^s at the point $\phi(m)$. This map, called the *differential* of ϕ, is denoted by $\phi_*: T_m(M^r) \to T_{\phi(m)}(N^s)$ and it is defined by

$$\phi_*(t)(f) = t(f \circ \phi)$$

for $t \in T_m(M^r)$ and $f \in \mathscr{F}_{\phi(m)}$.

With the notation of section 8.2, let γ_* be a tangent vector of $T_m(M^r)$ and let $\gamma: I \to M^r$ be the curve defining γ_*. The vector $\phi_*(\gamma_*) \in T_{\phi(m)}(N^s)$ is defined by the curve $\phi \circ \gamma: I \to N$. The map ϕ_* is clearly a homomorphism.

Let (U, u) be a chart of M^r containing m and let (V, v) be a chart of N^s containing $\phi(m)$. Let $\dfrac{du_1}{d\tau}, \ldots, \dfrac{du_r}{d\tau}$ be the components of γ_* with respect to the basis induced in $T_m(M^r)$ by the local coordinates at m. Let $\dfrac{dv_1}{d\tau}, \ldots, \dfrac{dv_s}{d\tau}$ be the components of $\phi_*\gamma_*$ with respect to the basis induced in $T_{\phi(m)}(N^s)$

by the local coordinates at $\phi(m)$. Then we have

$$\left(\frac{dv_i}{d\tau}\right) = \left(\frac{\partial v_i}{\partial u_j}\right)_m \left(\frac{du_j}{d\tau}\right),$$

where $i = 1, \ldots, s; j = 1, \ldots, r$, and $\left.\frac{\partial v_i}{\partial u_j}\right|_m = \left.\frac{\partial(v_i \circ \phi \circ u^{-1})}{\partial x_j}\right|_{x = u(m)}$.

12.2 Definition *The dimension of the vector space $\phi_*(T_m(M^r))$ is called the rank of the map $\phi: M^r \to N^s$ at m.*

With the above notation, the rank of ϕ at $m \in M^r$ is equal to the rank of the Jacobian matrix $\left(\frac{\partial v_i}{\partial u_j}\right)_m$.

12.3 Definition *Let M, N be two differentiable manifolds with* dim $M \leq$ dim N. *A differentiable map $\phi: M \to N$ is said to be regular at $m \in M$ if* rank ϕ = dim M.

The fact that $\phi: M \to N$ is regular at $m \in M$ is equivalent to the fact that the map $\phi_*: T_m(M) \to T_{\phi(m)}(N)$ is injective.

The map ϕ is called *regular* if it is regular at every point $m \in M$.

If $\phi: M \to N$ and $\psi: N \to P$ are two differentiable maps, the composition $\psi \circ \phi$ is defined. It is easy to check that the induced maps satisfy the condition:

$$(\psi \circ \phi)_* = \psi_* \circ \phi_*.$$

Let us consider again a differentiable map $\phi: M \to N$. To every differentiable function f defined on N is associated the differentiable function $f \circ \phi$ defined on M. We put

$$\phi^* f = f \circ \phi.$$

$\phi^* f$ is called the *inverse image* of f under ϕ. It is clear that:

$$\phi^*(f + g) = \phi^* f + \phi^* g,$$

$$\phi^*(fg) = (\phi^* f)(\phi^* g).$$

Let $m \in M$ and let $df \in T^*_{\phi(m)}(N)$. The map $\phi^*: T^*_{\phi(m)}(N) \to T^*_m(M)$, defined by $\phi^*(df) = d(\phi^* f)$, is a homomorphism which extends by linearity to a homomorphism

$$\phi^*: A^p_{\phi(m)} \to A^p_m$$

of the algebras of differential forms for every p.

Let (U, u) and (V, v) be local charts of M^r and N^s containing m and $\phi(m)$ respectively. If $\omega^p \in A^p_{\phi(m)}$, then:

$$\omega^p = \sum_{1 \leq i_1 < \cdots < i_p \leq s} \omega_{i_1 \cdots i_p} \, dv^{i_1} \wedge \cdots \wedge dv^{i_p},$$

and, for the inverse image of ω^p under ϕ, we have:

$$\phi^* \omega^p = \sum_{1 \leq i_1 < \cdots < i_p \leq s} \left(\sum_{i_1 \cdots i_p} \phi^* \omega_{i_1 \cdots i_p} \frac{\partial v_{i_1}}{\partial u_{j_1}} \cdots \frac{\partial v_{i_p}}{\partial u_{j_p}} \, du^{j_1} \wedge \cdots \wedge du^{j_p} \right).$$

The map ϕ^* has the following properties:

i) $\operatorname{supp} \phi^* \omega^p \subset \phi^{-1} (\operatorname{supp} \omega^p)$. In fact, if $m \notin \phi^{-1} (\operatorname{supp} \omega^p)$ then ω^p vanishes at $\phi(m)$ and, therefore, $\phi^* \omega^p$ vanishes at m;

ii) $\phi^* (\omega_1 \wedge \omega_2) = \phi^* \omega_1 \wedge \phi^* \omega_2$, which shows that ϕ^* is an algebra homomorphism;

iii) $\phi^* d\omega = d\phi^* \omega$ for every differential form ω; that is, ϕ^* commutes with the exterior differential.

We have seen that a differentiable map $\phi\colon M \to N$ induces two maps $\phi_*\colon T_m(M) \to T_{\phi(m)}(N)$ and $\phi^*\colon T^*_{\phi(m)}(N) \to T^*_m(M)$ of tangent and cotangent spaces at every $m \in M$. These two maps ϕ_* and ϕ^* are dual and they determine each other. From some properties of these maps we can get information on the map ϕ.

12.4 THEOREM *Let $\phi\colon M^r \to N^s$ be a differentiable map and let $m \in M^r$. Then we have:*

i) *If ϕ_* is surjective and $v_1, \ldots, v_s$ is any system of coordinates at $\phi(m)$, there exist $r - s$ functions $\psi_1, \ldots, \psi_{r-s}$ defined on a neighborhood of m such that $v_1 \circ \phi, \ldots, v_s \circ \phi, \psi_1, \ldots, \psi_{r-s}$ is a system of local coordinates at m.*

ii) *If ϕ_* is injective, we can choose r functions among $v_1 \circ \phi, \ldots, v_s \circ \phi$ which form a system of local coordinates at m. Moreover, to any system of coordinates $u_1, \ldots, u_r$ at m, there exists a system of coordinates $v'_1, \ldots, v'_s$ at $\phi(m)$ such that, in a neighborhood of m, $u_j = v'_j \circ \phi$ for $j = 1, \ldots, r$.*

iii) *If ϕ_* is an isomorphism, then ϕ defines a homeomorphism of a neighborhood of m onto a neighborhood of $\phi(m)$ whose inverse is differentiable at $\phi(m)$.*

Proof In case (i) we have $r \geq s$. Let $m \in M^r$ and let (U, u) be a local chart of M^r containing m. Then, for $i = 1, \ldots, s$, $v_i \circ \phi$ is a differentiable function of $u_1, \ldots, u_r$ on U. We shall show that rank $\left(\frac{\partial (v_i \circ \phi)}{\partial u_j}\right)_m = s$. In fact, assume that $\sum_{i=1}^{s} a_i \left(\frac{\partial (v_i \circ \phi)}{\partial u_j}\right)_m = 0$ for $1 \leq j \leq r$ and $a_i \in \mathbf{R}$. Let $t_i \in T_{\phi(m)}(N^s)$ be the vector defined by $t_i(v_k) = \delta_i^k$, $(1 \leq i, k \leq s)$. Since ϕ_* is surjective there is a vector $\tau_i \in T_m(M^r)$ such that $t_i = \phi_* \tau_i$. Then we have

$$\sum_{j=1}^{r} \left(\frac{\partial (v_k \circ \phi)}{\partial u_j}\right)_m \tau_i(u_j) = \tau_i (v_k \circ \phi) = \delta_i^k.$$

If we multiply by a_k and sum for $k = 1, \ldots, s$ we get $a_i = 0$.

If we assume, as we can, that the $s \times s$ matrix of the first s columns of $\left(\frac{\partial (v_i \circ \phi)}{\partial u_j}\right)_m$ has determinant different from zero, then we can take $\psi_1 = u_{s+1}, \ldots, \psi_{r-s} = u_r$, and $v_1 \circ \phi, \ldots, v_s \circ \phi, \psi_1, \ldots, \psi_{r-s}$ are a system of coordinates at $m \in M^r$.

In case (ii) we have $r \leq s$. Let $m \in M^r$ and let (U, u) be a local chart of M^r containing m. For $i = 1, \ldots, s$, $v_i \circ \phi$ is a differentiable function of $u_1, \ldots, u_r$ on U. We shall show that rank $\left(\frac{\partial (v_i \circ \phi)}{\partial u_j}\right)_m = r$. Assume that

$$\sum_{j=1}^{r} a^j \left(\frac{\partial (v_i \circ \phi)}{\partial u_j}\right) = 0 \quad \text{for} \quad 1 \leq i \leq s \quad \text{and} \quad a^j \in \mathbf{R}.$$

Let $\tau_j \in T_m(M^r)$ be the vector defined by $\tau_j(u_k) = \delta_j^k$, $(1 \leq j, k \leq r)$, and let $t = \sum_{j=1}^{r} a^j \tau_j$. Then we have

$$t\,(v_i \circ \phi) = \sum_{j=1}^{r} a^j \left(\frac{(\partial v_i \circ \phi)}{\partial u_j}\right)_m = 0,$$

or $\phi_* t(v_i) = 0$ for $1 \leq i \leq s$. It follows that $\phi_* t = 0$ and, since ϕ_* is injective, $t = 0$, which implies that $a^j = 0$ for $j = 1, \ldots, r$.

We can assume that the $r \times r$ matrix of the first r rows of $\left(\frac{\partial (v_i \circ \phi)}{\partial u_j}\right)_m$ has determinant different from zero. Then $v_1 \circ \phi, \ldots, v_r \circ \phi$ are a system of local coordinates on a neighborhood of $m \in M^r$ and in this neighborhood we have $u_j = \psi_j (v_1 \circ \phi, \ldots, v_r \circ \phi)$ with ψ_j differentiable functions. Then the functions:

$$v_j' = \psi_j (v_1, \ldots, v_r), \quad 1 \leq j \leq r,$$

$$v_j' = v_j, \qquad r + 1 \leq j \leq s,$$

form a system of local coordinates at $\phi(m) \in N^s$ and clearly $v_j' \circ \phi = u_j$ for $1 \leq j \leq r$.

(iii) follows at once from (i) and (ii). In fact, if ϕ_* is injective, there is a neighborhood V of m in M^r such that $\phi: V \to \phi(V)$ is a homeomorphism. If ϕ_* is surjective and V is any neighborhood of m in M^r, then $\phi(V)$ is a neighborhood of $\phi(m)$ in N^s. Moreover, when ϕ_* is an isomorphism, ϕ^{-1} is a differentiable map on $\phi(V)$. ○

13.1 Submanifolds

The notion of a submanifold generalizes that of a differentiable variety of E^n.

13.2 DEFINITION *Let M be a differentiable manifold. A differentiable manifold N is called a submanifold of M if there exists a regular injective map $i: N \to M$.*

If N is a submanifold of M, the map i is called an *embedding* and N is said to be *embedded* in M by i. We also have that $\dim N \leq \dim M$.

Let N be a submanifold of the manifold M and let $i: N \to M$ be the embedding. Let f be a differentiable function near $m \in i(N) \subset M$. Then, the function $f \circ i$, which is differentiable near $i^{-1}(m) \in N$, is called *the restriction* of f to N and is denoted by $f \mid N$.

Differentiable functions on N are locally restrictions of differentiable functions on M as is shown by the next theorem.

13.3 THEOREM *Let N be a submanifold of the differentiable manifold M. To every point $n \in N$ and differentiable function f on N there are a neighborhood U of n and a differentiable function g on M such that $g \circ i \mid U = f \mid U$.*

Proof It follows at once from (ii) of theorem 12.4. ○

We now give some examples of submanifolds.

13.4 *Open submanifolds* Any open subset U of a differentiable manifold M is a submanifold of M when the differentiable structure for U is given by the set of functions which are restrictions of differentiable functions on M to open subsets of U. The map $i: U \to M$ is the identity.

13.5 *Closed submanifolds* A submanifold N of M is a closed submanifold if $i(N)$ is a closed subset of M and every point $m \in i(N)$ is a contained in a local chart (U, u), where $u = (u_1, \ldots, u_n)$, with the property that $i(N) \cap U = \{u_{r+1} = \cdots = u_n = 0\}$.

If N is a submanifold of M, the topology of $i(N)$ as a subset of M is not necessarily the same as the topology of N. Thus the map $i\colon N \to M$, although injective and continuous, need not be a homeomorphism onto $i(N) \subset M$.

13.6 DEFINITION *A submanifold N of M is said to be regularly embedded if $i\colon N \to M$ is a homeomorphism into.*

13.7 THEOREM *A submanifold N is regularly embedded in M if and only if N is a closed submanifold of M.*

Proof It is trivial to prove that every closed submanifold is regularly embedded in an open submanifold of M. Therefore, we have only to prove the necessity.

Let $\dim N = r$ and $\dim M = n$. Let N be regularly embedded in M. Let $n \in N$ and let $u_1, \ldots, u_n$ be a system of coordinates at $i(n) \in M$. By theorem 12.4 there is a neighborhood of n where $u_1 \circ i, \ldots, u_r \circ i$ form a system of coordinates. The functions $u_h \circ i, h = r + 1, \ldots, n$ may be expressed in these coordinates as

$$u_h \circ i = f_h(u_1 \circ i, \ldots, u_r \circ i).$$

In a neighborhood of $i(n)$ we can replace the last $n - r$ coordinates by $\tilde{u}_h = u_h - f_h(u_1, \ldots, u_r), h = r + 1, \ldots, n$. We have then $\tilde{u}_h \circ i = 0$ for $h = r + 1, \ldots, n$.

Since i is a homeomorphism, there is a neighborhood $U_{i(n)}$ of $i(n)$ such that

$$U_{i(n)} \cap i(N) = \{\tilde{u}_{r+1} = \cdots = \tilde{u}_n = 0\}.$$

The manifold $M' = \bigcup_{i(n) \in i(N)} U_{i(n)}$ is an open submanifold of M which contains $i(N)$. We shall show now that $i(N)$ is relatively closed in M'. If a point $m \in M' \cap \overline{i(N)}$, then $m \in U_{i(n)}$ for some n. Since $U_{i(n)} \cap i(N)$ is relatively closed in $U_{i(n)}$, then $m \in i(N)$. Thus, $i(N)$ is a closed submanifold of M'. ○

13.8 *Remark* The above theorem shows that if (U, u) is a local chart of M such that U contains a point $m \in i(N)$, the functions $u_1 | i(N) \cap U, \ldots, u_r | i(N) \cap U$ are a system of local coordinates of $i(N)$. They define a differentiable structure on $i(N)$ which is equivalent to that of N although its definition does not depend on N. Thus, the theorem shows that if we consider a regular embedding, submanifold and image set have the same differentiable structure.

13.9 THEOREM *Every compact differentiable manifold can be embedded in an Euclidean space of sufficiently high dimension.*

Proof Let M be an n-dimensional compact differentiable manifold. By theorem 7.3 there are two finite coverings of M by coordinate neighborhoods, $\mathscr{U} = \{U_\alpha\}$, $\mathscr{V} = \{V_\alpha\}$, $\alpha = 1, \ldots, p$, such that for every α, $\bar{V}_\alpha \cap U_\alpha$.

Let (U_α, u_α) be a local chart of M. If $m \in U_\alpha$, then $u_{\alpha 1}(m), \ldots, u_{\alpha n}(m)$ are the coordinates of m.

For every $m \in M$ and every $\alpha = 1, \ldots, p$, let $f_\alpha(m)$ be a differentiable function such that $0 \leq f_\alpha(m) \leq 1$ and

$$f_\alpha(m) = \begin{cases} 1 & \text{if} \quad m \in \bar{V}_\alpha \\ 0 & \text{if} \quad m \in M \backslash U_\alpha . \end{cases}$$

For $j = 0, 1, \ldots, n$ and $\alpha = 1, \ldots, p$, define on M the functions

$$g_{\alpha j}(m) = \begin{cases} f_\alpha(m) & \text{if} \quad j = 0; \quad \alpha = 1, \ldots, p, \\ u_{\alpha j}(m) f_\alpha(m) & \text{if} \quad j = 1, \ldots, n; \quad \alpha = 1, \ldots, p; \quad m \in U_\alpha, \\ 0 & \text{if} \quad j = 1, \ldots, n; \quad \alpha = 1, \ldots, p; \quad m \in M \backslash U_\alpha . \end{cases}$$

The $(n + 1)p$ functions $g_{\alpha j}$ define a continuous map $i \colon M \to E^{(n+1)p}$. Since M is compact, $i(M)$ is also compact and thus closed.

We shall show that the map i is an embedding. Assume that $g_{\alpha j}(m) = g_{\alpha j}(m')$. Since for some α we have $g_{\alpha 0}(m) = 1$, we also have $g_{\alpha 0}(m') = 1$; hence, $m' \in U_\alpha$. Then for $j = 1, \ldots, n$ we have $u_{\alpha j}(m) = u_{\alpha j}(m')$ and thus $m = m'$, which proves that i is an injective map.

From the definition of the functions $g_{\alpha j}$, it follows at once that for every $m \in M$ the map

$$i_* : T_m(M) \to T_{i(m)}(E^{(n+1)p})$$

is also injective. ○

The above theorem is not the best possible embedding theorem. In fact, H. Whitney proved that any n-dimensional differentiable manifold can be embedded in an Euclidean space of dimension $2n + 1$.

13.10 DEFINITION *A non-injective regular map $i \colon N \to M$ is called an immersion if for every $n \neq n'$ such that $i(n) = i(n')$ we have $i_*(T_n(N)) \cap i_*(T_{n'}(N)) = 0 \in T_{i(n)}(M)$.*

We remark that, if a differentiable manifold N is immersed in a differentiable manifold M, then N is not a submanifold of M.

If N is a p-dimensional submanifold of the n-dimensional manifold M, the integer $n - p$ is called the *codimension* of N in M. A submanifold of codimension 1 is also called a *hypersurface.*

14.1 Integration

Let M be an n-dimensional orientable differentiable manifold and let $\mathscr{U} = \{U_\alpha\}$ be a locally finite open covering of M by coordinate neighborhoods. Assume that one of the two possible orientations for M is chosen.

Let ω be a differential form of degree n defined on M and let $\{\phi_\alpha\}$ be a C^∞ partition of unity subordinate to the covering $\mathscr{U}$. Then

$$\omega = \sum_\alpha \phi_\alpha \omega$$

where the differential form $\phi_\alpha \omega$ vanishes outside U_α. In the local chart (U_α, u_α) we have

$$\omega(m) = \omega_{1\ldots n}(m)\, du_\alpha^1 \wedge \cdots \wedge du_\alpha^n.$$

Let W be an open set relatively compact in M; we define

$$\int_{W \cap U_\alpha} \phi_\alpha \omega = \int_{u_\alpha(W \cap U_\alpha)} \phi_\alpha \omega_{1\ldots n}\, dx^1 \wedge \cdots \wedge dx^n$$

where the right-hand side is a well defined multiple integral on an open set of E^n.

Since W is relatively compact, we can define now

$$\int_W \omega = \sum_\alpha \int_{W \cap U_\alpha} \phi_\alpha \omega .$$

We must show that $\int_W \omega$ depends only on W and ω, that is, it does not depend on the covering $\{U_\alpha\}$ and on the partition of unity $\{\phi_\alpha\}$ subordinate to it. Let $\mathscr{V} = \{V_\beta\}$ be another locally finite open covering of M by coordinate neighborhoods. In the intersection $U \cap V$ of two coordinate neighborhoods $U \in \mathscr{U}$ and $V \in \mathscr{V}$, let $J = \det\left(\frac{\partial v_i}{\partial u_j}\right)$ be the Jacobian of the local coordinates $v_1, \ldots, v_n$ with respect to the local coordinates $u_1, \ldots, u_n$. From section 12.1 we know that

$$\omega = \omega_{1\ldots n}\, du^1 \wedge \cdots \wedge du^n = J^{-1} \omega_{1\ldots n}\, dv^1 \wedge \cdots \wedge dv^n.$$

On the other hand, the integral in E^n must be multiplied by $|J|$ and, since we can assume that for the chosen orientation of M we have $J > 0$, it follows that the integral $\int_W \omega$ is independent of the local coordinates. That the integral is independent of partitions of unity follows at once by the additivity property of integrals.

The number $\int_W \omega$ is called the *oriented integral* of the differential form ω over the open set $W \subset M$.

If the orientation of M is changed, then $|J| = -J$, and, if we denote by $-W$ the same open set W with the new orientation, we have:

$$\int_W \omega = -\int_{-W} \omega .$$

If W is not relatively compact in M, we can find a sequence of relatively compact open sets $\{W_i\}$ such that $W_i \subset W$ and $\lim_{i\to\infty} W_i = W$. Then we may define

$$\int_W \omega = \lim_{i\to\infty} \int_{W_i} \omega$$

if this limit exists and it is independent of the sequence $\{W_i\}$.

If supp ω is compact, $\int_W \omega$ is well defined for any open set $W \subset M$.

Let $W \subset M$ be an open set and let ∂W be its boundary. Let $p \in \partial W$, let U be an open neighborhood of p, and let f be a C^∞ function defined on U. We say that f defines ∂W at p if:

$$W \cap U = \{m \in U \mid f(m) < 0\},$$

$$(df)_p \neq 0 .$$

If to every point $p \in \partial W$ there are a coördinate neighborhood U and a function f with the above property, we say that *W is on one side of ∂W*.

The boundary of W is said to be *piecewise differentiable* if:

i) $\partial W = \Omega_1 \cup \cdots \cup \Omega_k \cup N$

where

Ω_i is an open subset of ∂W,

$\bar{\Omega}_i$ is a compact subset of an $(n-1)$-dimensional submanifold M_i regularly embedded in M,

N is a compact subset of a finite union of $(n-2)$-dimensional submanifolds of M,
for every $i \neq j$, $\bar{\Omega}_i \cap \bar{\Omega}_j \subset N$.

ii) W is on one side of ∂W on a neighborhood of every point of $\partial W \backslash N$.

If (U, u) is a local chart of M containing $p \in \Omega_i$ we can always assume that $\partial W \cap U = \{m \in U \mid u_n(m) < 0\}$. Then $(\partial W \cap U, u_1, \ldots, u_{n-1})$ is a local chart of Ω_i. Since M is oriented, the orientation of U induces an orientation on $\partial W \cap U$ and thus on Ω_i. This induces an orientation on ∂W and we shall always assume that ∂W is so oriented.

Let $\phi_i: \bar{\Omega}_i \to M$ be the embedding and let $\omega \in A^{n-1}(M)$. Then $\phi_i^* \omega$ is a differential form of degree $n-1$ on $\bar{\Omega}_i$. The integral

$$\int_{\Omega_i} \phi_i^* \omega$$

is well defined. We can therefore define

$$\int_{\partial W} \omega = \sum_{i=1}^{k} \int_{\Omega_i} \phi_i^* \omega .$$

By using partitions of unity, Stokes' theorem can be easily generalized to manifolds.

14.2 THEOREM *Let $W \subset M$ be an open set with piecewise differentiable boundary. Let ω be an $(n-1)$-differential form on M. Then*

$$\int_{\partial W} \omega = \int_{W} d\omega . \quad \bigcirc$$

We remark that Stokes' formula is valid either if supp ω is compact and W is any open set or if W is relatively compact in M and ω is any differential form.

15.1 Problems

1 (1.1) Let $f_1, \ldots, f_m$ be differentiable functions on an open set $\Omega \subset E^n$, $(m < n)$. Prove that the set $U = \left\{x \in \Omega \,\middle|\, \text{rank} \left(\frac{\partial f_i}{\partial x_j}\right) = m\right\}$ is open.

2 (1.1) Prove statements (a) and (b) of section 1.1.

3 (1.1) Show that the equation

$$(x^2 + y^2 + z^2 + 3)^2 - 16\,(x^2 + y^2) = 0$$

defines a 2-dimensional differentiable variety of E^3.

4 (1.1) Show that the set

$$S = \{(x, y, z) \in E^3 \mid x^2y^2 + x^2z^2 + y^2z^2 + 2xyz = 0\}$$

is not a differentiable variety of E^3. Find a set $K \subset S$ such that $S \backslash K$ is a 2-dimensional differentiable variety.

5 (3.1) Let $\mathscr{C}_1$, $\mathscr{C}_2$ be two bases for some differentiable structures on a topological space M. Prove that $\mathscr{C}_1$ and $\mathscr{C}_2$ are equivalent if and only if for every point $m \in M$ there are local charts $(U_1, u_1) \in \mathscr{C}_1$ and $(U_2, u_2) \in \mathscr{C}_2$ such that u_1 and u_2 are C^∞-related.

6 (3.1) Prove that every differentiable variety can be given the structure of a differentiable manifold.

7 (5.1) Prove that the projective space $\mathbf{P}^2(\mathbf{R})$ is a compact topological space.

8 (5.1) Prove that the projective space $\mathbf{P}^n(\mathbf{R})$ is an orientable manifold if and only if n is odd.

9 (6.1) Fill in the missing details in section 6.1.

10 (10.1) Prove that, for $X, Y \in \mathscr{V}(U)$ and $f, g \in \mathscr{D}(U)$,

$$[fX, gY] = fg\,[X, Y] + f(Xg)\,Y - g\,(Yf)\,X.$$

11 (11.1) Prove properties (i), (ii), and (iii) of the exterior differential stated in section 11.1.

12 (11.1) Prove that the definition of d satisfying the above properties is actually independent of the local coordinates.

13 (12.1) Show that the map $\phi\colon \mathbf{R} \to E^2$ given by $x = t^2$, $y = t^3$ is not regular.

14 (12.1) Prove properties (i), (ii), and (iii) of the map ϕ^* stated in section 12.1.

15 (13.1) Show that the map $i\colon \mathbf{R} \to E^2$ given by $x = t^2 - 1$, $y = t^3 - t$ is not injective but defines an immersion of $\mathbf{R}$ into E^2.

16 (13.1) Let N be a submanifold of M and let $\dim N = \dim M$. Show that N is open n M.

CHAPTER 4

Complex manifolds

1.1 Complex analytic manifolds

WE DENOTE by $\mathbf{C}^n$ the complex vector space of all n-tuples $z = (z_1, \ldots, z_n)$ of complex numbers where $z_j = x_j + iy_j$ with $i = \sqrt{-1}$ and $x_j, y_j \in \mathbf{R}$, $j = 1, \ldots, n$. The map

$$\varrho : \mathbf{C}^n \to V^{2n}$$

defined by $\varrho(z_1, \ldots, z_n) = (x_1, y_1, \ldots, x_n, y_n)$ identifies $\mathbf{C}^n$ with V^{2n}. If we consider $\mathbf{C}^n$ and V^{2n} as topological spaces, then ϱ is a homeomorphism.

We recall that a complex valued function f defined on an open set $U \subset \mathbf{C}^n$ is said to be *holomorphic* or *complex analytic* on U if f can be expanded into a convergent power series with positive radius of convergence at every $z^0 \in U$; that is, if

$$f(z) = \sum a_{i_1 \cdots i_n} (z_1 - z_1^0)^{i_1} \cdots (z_n - z_n^0)^{i_n}$$

for z in a suitable neighborhood of z^0 in U.

Let $U \subset \mathbf{C}^n$ and $V \subset \mathbf{C}^m$ be open sets. A map $\phi \colon U \to V$ is called a *holomorphic map*, or a *complex analytic map*, if, for every f holomorphic on V, the function $f \circ \phi$ is holomorphic on U. Since the map ϕ can be described by m functions $\phi_1, \ldots, \phi_m$, it follows that ϕ is a holomorphic map if and only if the m functions $\phi_1, \ldots, \phi_m$ are holomorphic on U.

Let U, V be open subsets of $\mathbf{C}^n$ and let $\alpha \colon U \to V$ be a homeomorphism. α is called a *holomorphic homeomorphism* if α is a holomorphic map. If α is a holomorphic homeomorphism, a function f defined on U is holomorphic on U if and only if f is the inverse image under α of a holomorphic function on V.

Let M be a Hausdorff topological space with countable topology. Let U be an open subset of M. A *coordinate system* for U is a homeomorphism u of U onto an open subset of $\mathbf{C}^n$. The pair (U, u) is a *local chart* of M. The complex valued functions $u_1, \ldots, u_n$ are *local coordinates* at every point $m \in U$.

Let (U, u) and (V, v) be two local charts of M. Then u, v are said to be *holomorphically related* if $u \circ v^{-1}$ and $v \circ u^{-1}$ are holomorphic maps wherever they are defined.

Let $\mathscr{C} = \{(U_\alpha, u_\alpha)\}$ be a set of local charts of M.

1.2 DEFINITION *An n-dimensional complex analytic manifold is a pair* $(M, \mathscr{C})$ *such that:*

i) $M = \cup_\alpha U_\alpha$,
ii) *for every* α, β *the maps* u_α, u_β *are holomorphically related,*
iii) $\mathscr{C}$ *is maximal with respect to* (i) *and* (ii).

If $(M, \mathscr{C})$ is a complex analytic manifold, $\mathscr{C}$ is called the *complex structure* of M. We shall write M for $(M, \mathscr{C})$ and simply say that M is a complex manifold. If M is a complex manifold of complex dimension n, the topological dimension of M is $2n$. We shall write $\dim_{\mathbf{C}} M = n$ and $\dim_{\mathbf{R}} M = 2n$.

1.3 THEOREM *Every complex manifold is an orientable real analytic manifold.*

Proof Let M be an n-dimensional complex manifold and let $\mathscr{C} = \{(U_\alpha, u_\alpha)\}$ be its complex structure. The composition of the u_α with the homeomorphism ϱ: $\mathbf{C}^n \to V^{2n}$ shows that $\mathscr{C}' = \{(U_\alpha, \varrho \circ u_\alpha)\}$ is a real analytic structure for M so that $(M, \mathscr{C}')$ is a $2n$-dimensional real analytic manifold. The fact that this real manifold is orientable and actually oriented follows easily from Cauchy–Riemann equations. ○

The above theorem shows two necessary conditions for a real analytic manifold M to have a complex structure. The first one is that $\dim_{\mathbf{R}} M$ be an even integer. The second one is that M be orientable. It can be proved that if $n = 1$, these conditions are also sufficient. We shall see later that if $n > 1$, these two conditions are not sufficient for M to be given a complex structure.

A connected 1-dimensional complex manifold is also called a Riemann surface.

If M is a complex manifold, the notion of holomorphic function on M is well defined. A complex valued function f defined on an open set $U \subset M$ is said to be holomorphic on U if every point $m \in U$ has a neighborhood where f can be expanded into a convergent power series of the local coordinates. That is, if m has coordinates $m_1, \ldots, m_n$ and (W, w) is a local chart containing m, then $f(m') = \Sigma a_{i_1 \cdots i_n} (w_1 - m_1)^{i_1} \cdots (w_n - m_n)^{i_n}$ for m' in a suitable neighborhood of m. The definition has an intrinsic meaning; that is, it does not depend on the local chart since, if (V, v) is another local chart, the coordinates $v_1, \ldots, v_n$ are holomorphic functions of the coordinates $w_1, \ldots, w_n$.

Antiholomorphic functions on M can also be defined. Let $w_1, \ldots, w_n$ be local coordinates at $m \in M$ and let $w_j = u_j + iv_j$, $j = 1, \ldots, n$. Then $\bar{w}_j = u_j - iv_j$ are the conjugate functions of the local coordinates w_j. A complex valued function f defined on an open set $U \subset M$ is said to be antiholomorphic on U if every point $m \in U$ has a neighborhood where f can be expanded into a convergent power series of the conjugates of the local coordinates; that is, if $f(m') = \Sigma\, a_{i_1 \cdots i_n} (\bar{w}_1 - \bar{m}_1)^{i_1} \cdots (\bar{w}_n - \bar{m}_n)^{i_n}$ for m' in a suitable neighborhood of m.

2.1 Examples

We give here some examples of complex manifolds leaving all the proofs to the reader.

2.2 *The topological space* $\mathbf{C}^n$ It is a complex manifold of complex dimension n which can be covered by the unique chart (U, u) where $U = \mathbf{C}^n$ and $u =$ identity.

2.3 *The Riemann surface* of an analytic function of one complex variable.

2.4 *The complex projective space* $\mathbf{P}^n(\mathbf{C})$ It is the quotient space of $\mathbf{C}^{n+1} \backslash \{0\}$ with respect to the equivalence relation

$$(z_1, \ldots, z_{n+1}) = (\lambda z_1, \ldots, \lambda z_{n+1})$$

where $\lambda \neq 0$ is any complex number. $\mathbf{P}^n(\mathbf{C})$ is an n-dimensional complex manifold which can be covered by the $n + 1$ charts (U_i, u_i) where

$$U_i = \{(z_1, \ldots, z_{n+1}) \in \mathbf{P}^n(\mathbf{C}) \mid z_i \neq 0\} \quad \text{and} \quad u_i : U_i \to \mathbf{C}^n$$

is the homeomorphism which to $(z_1, \ldots, z_{n+1}) \in \mathbf{P}^n(\mathbf{C})$ associates

$$\left(\frac{z_1}{z_i}, \ldots, \frac{z_{i-1}}{z_i}, \frac{z_{i+1}}{z_i}, \ldots, \frac{z_{n+1}}{z_i}\right) \in \mathbf{C}^n.$$

We remark that the real analytic structure of $\mathbf{P}^n(\mathbf{C})$ does not coincide with that of $\mathbf{P}^{2n}(\mathbf{R})$. In fact, $\mathbf{P}^n(\mathbf{C})$ as a real manifold is orientable, while $\mathbf{P}^{2n}(\mathbf{R})$ is not orientable.

2.5 More examples of complex manifolds such as product manifolds, complex tori, etc., may be produced using the same techniques as in 3.5.1.

3.1 Differentiable and complex structure of complex manifolds

Let M be a complex manifold of complex dimension n, let $m \in M$ and let (U, w) be a local chart of M containing m. A complex valued function $f = f' + if''$ defined on a neighborhood of m is said to be differentiable at m or C^∞ at m if the two functions f' and f'' are differentiable functions at m with respect to the differentiable structure of M as a real analytic manifold.

For $w_j = u_j + iv_j$ and $j = 1, \ldots, n$ we put

$$\frac{\partial f}{\partial u_j} = \frac{\partial f'}{\partial u_j} + i\frac{\partial f''}{\partial u_j}$$

$$\frac{\partial f}{\partial v_j} = \frac{\partial f'}{\partial v_j} + i\frac{\partial f''}{\partial v_j}.$$

A function f is differentiable on M if f is differentiable at every $m \in M$ with respect to the local coordinates.

If f is differentiable on M, we define, in every local chart (U, w),

$$\frac{\partial f}{\partial w_j} = \frac{1}{2}\left(\frac{\partial f}{\partial u_j} - i\frac{\partial f}{\partial v_j}\right),$$

$$\frac{\partial f}{\partial \bar{w}_j} = \frac{1}{2}\left(\frac{\partial f}{\partial u_j} + i\frac{\partial f}{\partial v_j}\right).$$

These derivatives may be used to characterize holomorphic and antiholomorphic functions on M. This is stated in the next theorem, the proof of which we leave to the reader.

3.2 THEOREM *A differentiable function f defined on an open set $U \subset M$ is holomorphic on U if and only if*

$$\frac{\partial f}{\partial \bar{w}_j} = 0, \quad j = 1, \ldots, n,$$

on U. The function f is antiholomorphic on U if and only if

$$\frac{\partial f}{\partial w_j} = 0, \quad j = 1, \ldots, n,$$

on U. ○

We shall now show that the maximum modulus principle holds for holomorphic functions of several complex variables defined on a complex manifold. If f is a holomorphic function on an open set U, then $|f|$ attains its maximum at some points on the boundary ∂U of U.

3.3 THEOREM *Let U be a connected open set relatively compact in a complex manifold M. Let f be a holomorphic function on an open neighborhood of $U \cup \partial U$. Then, for every $m \in U$,*

$$|f(m)| \leq \max_{p \in \partial U} |f(p)|.$$

Proof If f is a constant, the theorem is trivial and the equality holds. Let f be a non-constant holomorphic function on an open neighborhood of $U \cup \partial U$. Since $U \cup \partial U$ is compact, there exists a point $m_0 \in U \cup \partial U$ where the real valued function $|f|$ attains its maximum. Assume $m_0 \in U$. Let (W, w) be a local chart of M containing m_0 such that $W \subset U$. Consider the function $g = f \circ w^{-1}$ defined on $w(W)$. Apply now the maximum modulus principle to the functions of one complex variable, restrictions of g to all complex lines through $z_0 = w(m_0)$. Since not all these functions can be constant, there are points $z \in \overline{w(W)}$ such that $|g(z)| > |g(z_0)|$.

Let $m = w^{-1}(z)$. We have $|f(m)| > |f(m_0)|$ and, since $m \in U \cup \partial U$, this inequality contradicts the assumption that $|f(m_0)| = \max |f|$. ○

3.4 COROLLARY *Let M be a compact complex manifold. The only global holomorphic functions on M are the constants.*

Proof In the above theorem we can take $U = M$. Since $\partial U = \emptyset$, for every f holomorphic on M the max $|f|$ is taken at some point $m \in M$. This contradicts the theorem if f is not constant. ○

3.5 *Remark* That 1-dimensional complex manifolds have properties which do not generalize to higher dimensional manifolds is also shown by the following example. If M is a non-compact complex manifold and $\dim_{\mathbf{C}} M = 1$, there exist non-constant holomorphic functions on M. We shall show that if $\dim_{\mathbf{C}} M > 1$ this is not necessarily true. In fact, let M be a compact, connected, complex manifold of complex dimension >1. Let $m \in M$, then $N = M \setminus \{m\}$ is non-compact. If there were a non-constant holomorphic function f on N, then f would be holomorphic also at m by Hartog's theorem. But then f would be constant by corollary 3.4.

Germs of holomorphic functions on a complex manifold may be defined following the lines of section 3.6.1.

Let M be a complex manifold and let $m \in M$. A function f will be called holomorphic *near m* if f is holomorphic on some open neighborhood of m. Two functions f and g, holomorphic near m, will be called equivalent if they have the same value on an open neighborhood m; that is, if f is holomorphic on an open neighborhood U of m and g is holomorphic on an open neighborhood V of m, we say that f and g are equivalent if there exists an open set W such that $m \in W \subset U \cap V$ and $f \mid W = g \mid W$. It follows then that $f \mid U \cap V = g \mid U \cap V$. Each equivalence class of functions will be called a *germ of holomorphic function at m.* The set of all germs at $m \in M$ will be denoted by $\mathcal{O}_m$. The germ of f will be denoted by $\mathscr{f}$.

If f_1, f_2 are two representatives of the germs $\mathscr{f}_1, \mathscr{f}_2 \in \mathcal{O}_m$, the sum $\mathscr{f}_1 + \mathscr{f}_2$ is defined as the germ of the function $f_1 + f_2$ and the product $\mathscr{f}_1 \mathscr{f}_2$ is defined as the germ of the function $f_1 f_2$. The functions $f_1 + f_2$ and $f_1 f_2$ are clearly holomorphic on some neighborhood of m and it is easy to check that the definitions of $\mathscr{f}_1 + \mathscr{f}_2$ and $\mathscr{f}_1 \mathscr{f}_2$ do not depend on the choice of f_1 and f_2.

These two operations make $\mathcal{O}_m$ into a commutative ring with identity. The zero of the ring is the germ of the function which is identically zero. The identity element is the germ of the function which is identically one.

3.6 THEOREM *The ring $\mathcal{O}_m$ of germs of holomorphic functions at $m \in M$ is isomorphic to the ring of convergent power series about m.*

Proof Every power series about m which converges on some neighborhood of m represents a holomorphic function on that neighborhood and thus determines a unique element of $\mathcal{O}_m$.

Conversely, if $\mathscr{f} \in \mathcal{O}_m$, let f be a representative of $\mathscr{f}$. Then f has a power series expansion which converges on some neighborhood of m. Any other representative of $\mathscr{f}$ coincides with f on some neighborhood of m and thus has the same power series expansion.

Moreover, the operations on the two rings are pointwise on representatives and thus they are the same. ○

3.7 THEOREM *Let m, m' be two points of the complex manifold M. Then $\mathcal{O}_m$ is isomorphic to $\mathcal{O}_{m'}$.*

The proof is the same as in 3.6.2. ○

3.8 THEOREM *The ring $\mathcal{O}_m$ is an integral domain.*

Proof Let $\mathscr{f}, \mathscr{g} \in \mathcal{O}_m$ such that $\mathscr{f}\mathscr{g} = 0$ and let f, g be two representatives of $\mathscr{f}$ and $\mathscr{g}$ respectively. Then f, g are holomorphic functions on some open neighborhood U of m and we can assume that (U, w) is a local chart of M.

Since $fg = 0$, we have $f(w)\,g(w) = 0$ identically for w in some open neighborhood $W \subset U$. Assume that $f(w) \neq 0$ at some point $w \in W$. Since f is continuous, there is a neighborhood of w where $f \neq 0$ and, therefore, in that neighborhood $g(w) = 0$ identically. It follows that $g = 0$ identically in W and thus $g = 0$. ○

Let M, N be complex manifolds, let $\dim_{\mathbf{C}} M = p$, $\dim_{\mathbf{C}} N = q$ and let U be an open subset of M.

3.9 Definition *A map $\phi: U \to N$ is called a holomorphic map if for every $m \in U$ and holomorphic function f near $\phi(m)$ the function $f \circ \phi$ is holomorphic near m.*

If $\phi = (\phi_1, \ldots, \phi_q)$, then ϕ is a holomorphic map if and only if the functions $\phi_1, \ldots, \phi_q$ are holomorphic on U. It turns out that ϕ is a holomorphic map if and only if the local coordinates on every chart of $\phi(U) \subset N$ are holomorphic functions of the local coordinates on $U \subset M$.

A holomorphic map $\phi: M \to N$, which is a homeomorphism and such that ϕ^{-1} is holomorphic is called a *biholomorphic map*.

4.1 Holomorphic and antiholomorphic tangent vectors

Let M be a complex manifold, $\dim_{\mathbf{C}} M = n$. Let $m \in M$ and let $\mathscr{F}_m$ be the ring of germs of complex valued differentiable functions at m. $\mathscr{F}_m$ contains the ring $\mathcal{O}_m$ of germs of holomorphic functions at m and the ring $\bar{\mathcal{O}}_m$ of germs of antiholomorphic functions at m.

Let $\tau_m(M)$ be the vector space of all complex linear functionals

$$t: \mathscr{F}_m \to \mathbf{C}$$

which are derivations on $\mathscr{F}_m$.

Define

$$H_m(M) = \{t \in \tau_m(M) \mid \langle f, t\rangle = 0 \quad \text{if} \quad f \in \bar{\mathcal{O}}_m\},$$

$$\bar{H}_m(M) = \{t \in \tau_m(M) \mid \langle f, t\rangle = 0 \quad \text{if} \quad f \in \mathcal{O}_m\}.$$

Elements of $H_m(M)$ are called *holomorphic tangent vectors* to M at m and elements of $\bar{H}_m(M)$ are called *antiholomorphic tangent vectors* to M at m. Both H_m and $\bar{H}_m$ are complex vector spaces of dimension n. H_m is called the *holomorphic tangent space* to M at m, and $\bar{H}_m$ the *antiholomorphic tangent space* to M at m.

The vector space H_m may also be described as follows: Let $\Delta = \{\lambda \in \mathbf{C} \mid |\lambda| < 1\}$ be the unit disc in $\mathbf{C}$. Consider the set of all holomorphic maps

$\gamma: \Delta \to M$ such that $\gamma(0) = m$. Let (U, w) be a local chart of M containing m and let γ_1, γ_2 be two such maps. Let γ_1 be given by $w_j = w_j(\lambda_1), j = 1, \ldots, n$ and γ_2 by $w_j = w_j(\lambda_2), j = 1, \ldots, n$. We identify γ_1 and γ_2 if

$$\left(\frac{dw_j}{d\lambda_1}\right)_{\lambda_1=0} = \left(\frac{dw_j}{d\lambda_2}\right)_{\lambda_2=0}$$

for $j = 1, \ldots, n$.

Each equivalence class of holomorphic maps γ is then an element of H_m by defining

$$\gamma_*(\lambda)(f) = \left(\frac{d(f \circ \gamma)}{d\lambda}\right)_{\lambda=0}.$$

Since $\bar{H}_m$ is the complex conjugate space of H_m, its elements may as well be defined as equivalence classes of antiholomorphic maps $\bar{\gamma}: \Delta \to M$ by identifying two maps $\bar{\gamma}_1, \bar{\gamma}_2$ when, with obvious meaning of symbols,

$$\left(\frac{d\bar{w}_j}{d\lambda_1}\right)_{\lambda_1=0} = \left(\frac{d\bar{w}_j}{d\lambda_2}\right)_{\lambda_2=0}$$

for $j = 1, \ldots, n$.

The next theorem follows at once from the definitions.

4.2 THEOREM *For every $m \in M$, $H_m \cap \bar{H}_m = 0 \in \tau_m$ and $H_m \oplus \bar{H}_m = \tau_m$.* ○

Let $\gamma_*(\lambda) \in H_m$ and let (U, w) be a local chart of M containing $m = \gamma(0)$. Consider the map $\gamma_*(\lambda) \to \left(\frac{dw_1}{d\lambda}, \ldots, \frac{dw_n}{d\lambda}\right)$ which, to the holomorphic vector $\gamma_*(\lambda)$, associates the n-tuple of its components. If (V, z) is another local chart with $m \in U \cap V$, the same vector $\gamma_*(\lambda)$ also has components

$$\left(\frac{dz_1}{d\lambda}, \ldots, \frac{dz_n}{d\lambda}\right).$$

Then we have

$$\left(\frac{dz_j}{d\lambda}\right) = \left(\frac{\partial z_j}{\partial w_h}\right)_m \left(\frac{dw_h}{d\lambda}\right).$$

In the same way, if $\bar{\gamma}_*(\lambda)$ is an antiholomorphic tangent vector whose components are $\left(\frac{d\bar{w}_1}{d\lambda}, \ldots, \frac{d\bar{w}_n}{d\lambda}\right)$ and $\left(\frac{d\bar{z}_1}{d\lambda}, \ldots, \frac{d\bar{z}_n}{d\lambda}\right)$, we have

$$\left(\frac{d\bar{z}_j}{d\lambda}\right) = \left(\overline{\frac{\partial z_j}{\partial w_h}}\right)_m \left(\frac{d\bar{w}_h}{d\lambda}\right).$$

We have seen that, if M is an n-dimensional complex manifold, then M is also a $2n$-dimensional real analytic manifold. If $m \in M$, (U, w) is a local chart of the complex structure of M containing m, and $w = (w_1, \ldots, w_n)$ with $w_j = u_j + iv_j$, $j = 1, \ldots, n$, then $(U, \varrho \circ w)$ is a local chart of the real structure of M. With respect to this structure, the tangent space T_m to M at m is also defined. T_m is a $2n$-dimensional real vector space; a basis of T_m is $\left\{\frac{\partial}{\partial u_1}, \ldots, \frac{\partial}{\partial u_n}, \frac{\partial}{\partial v_1}, \ldots, \frac{\partial}{\partial v_n}\right\}$. Every element of T_m is a linear combination with real coefficients of these vectors.

We may now consider the complexification of T_m, that is, the vector space $\tilde{T}_m = \mathbf{C} \otimes T_m$. Every element of $\tilde{T}_m$ is a linear combination with complex coefficients of the same vectors of the basis of T_m. It is easy to prove that

$$\tilde{T}_m = T_m \oplus iT_m.$$

At every point $m \in M$ we have then defined five tangent spaces, $T_m, \tilde{T}_m, H_m, \bar{H}_m$, and τ_m, closely related to each other.

A basis of H_m is given by $\left\{\frac{\partial}{\partial w_1}, \ldots, \frac{\partial}{\partial w_n}\right\}$ and a basis of $\bar{H}_m$ is given by $\left\{\frac{\partial}{\partial \bar{w}_1}, \ldots, \frac{\partial}{\partial \bar{w}_n}\right\}$. Thus, $\left\{\frac{\partial}{\partial w_1}, \ldots, \frac{\partial}{\partial w_n}, \frac{\partial}{\partial \bar{w}_1}, \ldots, \frac{\partial}{\partial \bar{w}_n}\right\}$ is a basis of τ_m. All these are complex vector spaces, and, clearly, τ_m is isomorphic to $\tilde{T}_m$.

Let U be an open set of M. A contravariant vector field $L = \{L_m\}$ on U is a function which, to every $m \in U$, associates a vector $L_m \in \tau_m$. The assignment of L is equivalent to the assignment of n complex valued functions L^i on U. L is called differentiable on U if the L^i are differentiable on U, that is, if in every local chart the L^i are differentiable functions with respect to the real structure of M.

4.3 DEFINITION *A vector field L on U is called a holomorphic vector field if,*

i) *to every $m \in U$, L assigns a vector $L_m \in H_m$,*
ii) *the L^i are holomorphic functions on U.*

The definition of an *antiholomorphic vector field* is similar.

5.1 Holomorphic and antiholomorphic cotangent vectors

The vector space H_m^* dual of the holomorphic tangent space H_m is called the *holomorphic cotangent space* to M at m. The vector space $\bar{H}_m^*$ dual of the antiholomorphic tangent space $\bar{H}_m$ is called the *antiholomorphic cotangent space* to M at m. H_m^* and $\bar{H}_m^*$ are both n-dimensonal complex vector spaces. Their elements are called *holomorphic* and *antiholomorphic cotangent vectors* respectively.

Bases of H_m^* and $\bar{H}_m^*$ dual of the bases $\left\{\frac{\partial}{\partial w_j}\right\}$ and $\left\{\frac{\partial}{\partial \bar{w}_j}\right\}$ of H_m, $\bar{H}_m$, induced by the same system of coordinates at m, are given by $\{dw^j\}$ and $\{d\bar{w}^j\}$ respectively. Then for every $t^* \in H_m^*$ we have

$$t^* = \sum_{j=1}^{n} t_j^* \, dw^j$$

with $t_j^* = \left\langle \frac{\partial}{\partial w_j}, t^* \right\rangle$, and for every $\bar{t}^* \in \bar{H}_m^*$ we have

$$\bar{t}^* = \sum_{j=1}^{n} \bar{t}_j^* \, d\bar{w}^j$$

with $\bar{t}_j^* = \left\langle \frac{\partial}{\partial \bar{w}_j}, \bar{t}^* \right\rangle$.

If now $f \in \mathscr{F}_m$, there are two maps

$$\partial : \mathscr{F}_m \to H_m^*$$

defined by

$$\langle t, \partial f \rangle = \langle f, t \rangle, \quad t \in H_m,$$

and

$$\bar{\partial} : \mathscr{F}_m \to \bar{H}_m^*$$

defined by

$$\langle \bar{t}, \bar{\partial} f \rangle = \langle f, \bar{t} \rangle, \quad \bar{t} \in \bar{H}_m.$$

In particular, for $t = \frac{\partial}{\partial w_j}$, $f = w_h$ we have

$$\left\langle \frac{\partial}{\partial w_j}, \quad \partial w_h \right\rangle = \delta_j^h,$$

and for $\bar{t} = \frac{\partial}{\partial \bar{w}_j}$, $f = w_h$ we have

$$\left\langle \frac{\partial}{\partial \bar{w}_j}, \quad \bar{\partial} \bar{w}_h \right\rangle = \delta_j^h.$$

Hence, for $f \in \mathscr{F}_m$ we have

$$\partial f = \sum_{j=1}^{n} \frac{\partial f}{\partial w_j} \partial w_j$$

and

$$\bar{\partial} f = \sum_{j=1}^{n} \frac{\partial f}{\partial \bar{w}_j} \bar{\partial} \bar{w}_j .$$

∂f *is called the holomorphic differential of* f and $\bar{\partial} f$ *is called the antiholomorphic differential of* f.

If (U, w) and (V, z) are two local charts of M containing m, then we have

$$\left(\frac{\partial f}{\partial w_j}\right) = {}^t\!\left(\frac{\partial z_h}{\partial w_j}\right)_m \left(\frac{\partial f}{\partial z_h}\right)$$

and

$$\left(\frac{\partial f}{\partial \bar{w}_j}\right) = {}^t\!\left(\overline{\frac{\partial z_h}{\partial w_j}}\right)_m \left(\frac{\partial f}{\partial \bar{z}_h}\right).$$

We put $\tau_m^* = H_m^* \oplus \bar{H}_m^*$. It is clear, in fact, that τ_m^* is the dual of the complex tangent space τ_m.

For every $f \in \mathscr{F}_m$ we put

$$df = \partial f + \bar{\partial} f.$$

The element $df \in \tau_m^*$ is called the *total differential* of f. We remark that, if $f \in \mathcal{O}_m$, then $df = \partial f$ and, if $f \in \bar{\mathcal{O}}_m$, then $df = \bar{\partial} f$. Thus, the symbol df can always be used to denote the differential of such a function; in particular, have $\partial f = \sum_{j=1}^{n} \frac{\partial f}{\partial w_j} dw^j$ and $\bar{\partial} f = \sum_{j=1}^{n} \frac{\partial f}{\partial \bar{w}_j} d\bar{w}^j$.

5.2 THEOREM *The complex cotangent space* τ_m^* *is the complexification of the real cotangent space* T_m^*; *that is,* $\tau_m^* = \mathbf{C} \otimes T_m^*$.

Proof In a local chart (U, w) of M containing m, a basis of τ_m^* is given by $\{dw^1, \dots, dw^n, d\bar{w}^1, \dots, d\bar{w}^n\}$. Then every element of τ_m^* is a linear combination with complex coefficients of these differentials.

If $w_j = u_j + iv_j, j = 1, \dots, n$, the functions

$$u_j = \tfrac{1}{2}(w_j + \bar{w}_j)$$

$$v_j = \frac{1}{2i}(w_j - \bar{w}_j)$$

are, as we know, in $\mathscr{F}_m$. Their differentials

$$du^j = \tfrac{1}{2}(dw^j + d\bar{w}^j)$$

$$dv^j = \frac{1}{2i}(dw^j - d\bar{w}^j)$$

are elements of τ_m^*. By solving these equations we get

$$dw^j = du^j + idv^j$$

$$d\bar{w}^j = du^j - idv^j,$$

which shows that $\{du^1, \ldots, du^n, dv^1, \ldots, dv^n\}$ is another basis of τ_m^*.

If we now consider M as a real analytic manifold, we may consider the cotangent space T_m^*. Since a basis of T_m^* in the local chart $(U, \varrho \circ w)$ is given by $\{du^1, \ldots, du^n, dv^1, \ldots, dv^n\}$, it follows that every element of T_m^* is a linear combination with real coefficients of these differentials. ○

It follows from the above theorem that, if $f = f' + if''$, then $df = df' + idf''$ and df', df'' may be thought to be complex or real differentials as well.

6.1 Differential forms

Let M be an n-dimensional complex manifold and let $m \in M$. We construct the Grassmann algebra over $\mathbf{C}$ of the vector space τ_m^* and put $E_m^p = \wedge^p \tau_m^*$. Then we have

$$\wedge\tau_m^* = E_m = E_m^0 \oplus E_m^1 \oplus \cdots \oplus E_m^{2n}$$

where $E_m^0 = \mathbf{C}$ and $E_m^1 = \tau_m^*$.

If we recall now that $\tau_m^* = H_m^* \oplus \bar{H}_m^*$ and put $E^{r,s} = (\wedge^r H_m^*) \wedge (\wedge^s \bar{H}_m^*)$, then we have

$$E_m^p = \bigoplus_{r+s=p} E^{r,s}.$$

If (U, w) is a local chart containing m, the vector space $E_m^{r,s}$ has the natural basis

$$\{dw^{i_1} \wedge \cdots \wedge dw^{i_r} \wedge d\bar{w}^{j_1} \wedge \cdots \wedge d\bar{w}^{j_s}\}_{\substack{1 \le i_1 < \cdots < i_r \le n \\ 1 \le j_1 < \cdots < j_s \le n}}.$$

The union of all elements of the bases of all $E_m^{r,s}$ for $r + s = p$ is a basis of E_m^p, which is called the complex basis. According to theorem 5.2, E_m^p also has the real basis obtained by taking all exterior products of p elements in $(du^1, \ldots, du^n, dv^1, \ldots, dv^n)$.

In particular, the elements

$$dw^1 \wedge \cdots \wedge dw^n \wedge d\bar{w}^1 \wedge \cdots \wedge d\bar{w}^n$$

and

$$du^1 \wedge \cdots \wedge du^n \wedge dv^1 \wedge \cdots \wedge dv^n$$

are a complex and real basis of $E^{2n} = E^{n,n}$. A straightforward computation shows that

$$du^1 \wedge \cdots \wedge du^n \wedge dv^1 \wedge \cdots \wedge dv^n = (i/2)^n \, dw^1 \wedge \cdots \wedge dw^n \wedge d\bar{w}^1 \wedge \cdots \wedge d\bar{w}^n.$$

An element $\omega \in E^p_m$ is called *real* if its coefficients with respect to a real basis of E^p_m are real.

With respect to the complex basis of E^p_m, an element $\omega(m) \in E^p_m$ has a representation as a linear combination with complex coefficients of elements of that basis. For instance, the component of ω of type (r, s), $r + s = p$ is written as

$$\omega^{r,s}(m) = \sum_{\substack{1 \le i_1 < \cdots < i_r \le n \\ 1 \le j_1 < \cdots < j_s \le n}} \omega_{i_1 \cdots i_r j_1 \cdots j_s} \, dw^{i_1} \wedge \cdots \wedge dw^{i_r} \wedge d\bar{w}^{j_1} \wedge \cdots \wedge d\bar{w}^{j_s}.$$

We define the conjugate $\bar{\omega}$ of ω to be that element of E^p_m obtained from ω by replacing every dw^j with $d\bar{w}^j$ and vice versa and every coefficient with its complex conjugate. Then the component $\omega^{r,s}(m)$ of $\omega(m)$ yields the component of type (s, r) of $\overline{\omega(m)}$ as

$$\overline{\omega^{r,s}(m)} = \sum_{\substack{1 \le i_1 < \cdots < i_r \le n \\ 1 \le j_1 < \cdots < j_s \le n}} \overline{\omega_{i_1 \cdots i_r j_1 \cdots j_s}} \, d\bar{w}^{i_1} \wedge \cdots \wedge d\bar{w}^{i_r} \wedge d\bar{w}^{j_1} \wedge \cdots \wedge dw^{j_s}.$$

It is clear that an element $\omega \in E^p_m$ is real if and only if $\omega = \bar{\omega}$.

If $\omega \in E^p_m$ by putting $\xi = \frac{1}{2}(\omega + \bar{\omega})$ and $\eta = \dfrac{1}{2i}(\omega - \bar{\omega})$, we get

$$\omega = \xi + i\eta.$$

Then every element $\omega \in E^p_m$ is of this form with ξ, η real forms and, since this decomposition does not depend on the chosen local chart, we have:

6.2 THEOREM *The complex vector space* $E^p_m = \wedge^p \tau^*_m$ *is the complexification of the real vector space* $A^p_m = \wedge^p T^*_m$, *that is,* $E^p_m = \mathbf{C} \otimes A^p_m$. ○

Let W be an open subset of M. A function which, to every $m \in W$, associates an element $\omega^p(m) \in E^p_m$ such that the coefficients of ω in every local chart are complex valued differentiable functions is called a *differential form of degree p on W*.

If, for every $m \in W$, $\omega^p(m) \in E^{r,s}_m$, the differential form is said to be *homogeneous of type* (r, s) and is denoted by $\omega^{r,s}$.

If (U, w) and (V, z) are two local charts of M containing m, the transformation law for the coefficients of ω follows easily from that of differentials as in section 2.11.1.

We remark that, if ω is of type (p, o), the coefficients transform by holomorphic functions and if ω is of type (o, p) they transform by antiholomorphic functions.

6.3 DEFINITION *A differential form ω of type (p, o) is called holomorphic if in every local chart the coefficients of ω are holomorphic functions. A differential form of type (o, p) is called antiholomorphic if in every local chart its coefficients are antiholomorphic functions.*

If W is an open subset of M, we shall denote by $E^{r,s}(W)$, $E^p(W)$ the vector spaces of differential forms of type (r, s) and of degree p defined on W. An element $\omega^{r,s} \in E^{r,s}(W)$ has local expression

$$\omega^{r,s} = \sum_{\substack{1 \leq i_1 < \cdots < i_r \leq n \\ 1 \leq j_1 < \cdots < j_s \leq n}} \omega_{i_1 \cdots i_r j_1 \cdots j_s} \, dw^{i_1} \wedge \cdots \wedge dw^{i_r} \wedge d\bar{w}^{j_1} \wedge \cdots \wedge d\bar{w}^{j_s}.$$

We define two bihomogeneous exterior differentials

$$\partial : E^{r,s}(W) \to E^{r+1,s}(W)$$

$$\bar{\partial} : E^{r,s}(W) \to E^{r,s+1}(W)$$

by

$$\partial \omega^{r,s} = \sum_{\substack{1 \leq i_1 < \cdots < i_r \leq n \\ 1 \leq j_1 < \cdots < j_s \leq n}} (\partial \omega_{i_1 \cdots i_r j_1 \cdots j_s}) \wedge dw^{i_1} \wedge \cdots \wedge dw^{i_r} \wedge d\bar{w}^{j_1} \wedge \cdots \wedge d\bar{w}^{j_s}$$

and by

$$\bar{\partial} \omega^{r,s} = \sum_{\substack{1 \leq i_1 < \cdots < i_r \leq n \\ 1 \leq j_1 < \cdots < j_s \leq n}} (\bar{\partial} \omega_{i_1 \cdots i_r j_1 \cdots j_s}) \wedge dw^{i_1} \wedge \cdots \wedge dw^{i_r} \wedge d\bar{w}^{j_1} \wedge \cdots \wedge d\bar{w}^{j_s}.$$

If ω is of type (o, o), then $\partial \omega$ is the holomorphic differential of the function ω and $\bar{\partial} \omega$ is the antiholomorphic differential of ω.

We then define

$$d = \partial + \bar{\partial} : E^p(W) \to E^{p+1}(W).$$

The differential d is not bihomogeneous; however, it follows from its definition that, if ω is of type (r, s), then $d\omega$ is the sum of two differential forms of types $(r + 1, s)$ and $(r, s + 1)$.

Since every differential form ω is of the form $\xi + i\eta$ with ξ, η real differential forms, we have

$$d\omega = d\xi + i \, d\eta,$$

where $d\xi$, $d\eta$ may be thought to be real or complex differential forms as well. It follows that, as in the real case, for every differential form ω we have

$$d^2\omega = 0.$$

Since $d = \partial + \bar{\partial}$, this implies $\partial^2\omega + (\partial\bar{\partial} + \bar{\partial}\partial)\,\omega + \bar{\partial}^2\omega = 0$. But these forms are all of different types and thus we have:

$$\partial^2 = 0, \quad \bar{\partial}^2 = 0, \quad \partial\bar{\partial} + \bar{\partial}\partial = 0.$$

6.4 DEFINITION *A differential form ω is called ∂-closed, $\bar{\partial}$-closed, d-closed if $\partial\omega = 0$, $\bar{\partial}\omega = 0$, $d\omega = 0$, respectively. A differential form ω of type (r, s) is called ∂-exact, $\bar{\partial}$-exact, d-exact if there exists a form η of type $(r - 1, s)$ such that $\partial\eta = \omega$, η of type $(r, s - 1)$ such that $\bar{\partial}\eta = \omega$, η of degree $r + s - 1$ such that $d\eta = \omega$ respectively.*

Since $\partial^2 = 0$, every ∂-exact differential form is ∂-closed; since $\bar{\partial}^2 = 0$, every $\bar{\partial}$-exact form is $\bar{\partial}$-closed and, since $d^2 = 0$, every d-exact form is d-closed.

Theorem 3.2 can be easily generalized to differential forms.

6.5 THEOREM *A differential form $\omega^{p,0}$ defined on an open set $U \subset M$ is holomorphic on U if and only if*

$$\bar{\partial}\omega^{p,0} = 0$$

on U. A differential form $\omega^{0,p}$ is antiholomorphic on U if and only if

$$\partial\omega^{0,p} = 0$$

on U. ○

7.1 Action of maps

Let M, N be two complex manifolds and let $\phi\colon M \to N$ be a holomorphic map. As in the real case, ϕ induces homomorphisms between tangent and cotangent spaces, which we shall describe briefly, leaving all details to the reader. Let $m \in M$.

The holomorphic differential of ϕ is the homomorphism

$$\phi_* : H_m(M) \to H_{\phi(m)}(N)$$

defined by

$$\phi_*(t)(f) = t(f \circ \phi)$$

for $t \in H_m(M)$ and $f \in \mathcal{O}_{\phi(m)}$.

The antiholomorphic differential of ϕ is the homomorphism

$$\bar{\phi}_* : \bar{H}_m(M) \to \bar{H}_{\phi(m)}(N)$$

defined by

$$\bar{\phi}_*(\bar{t})(f) = \bar{t}(f \circ \bar{\phi})$$

for $\bar{t} \in \bar{H}_m(M)$ and $f \in \bar{\mathcal{O}}_{\phi(m)}$. The antiholomorphic differential of ϕ is the differential of the map $\bar{\phi}$.

The holomorphic map ϕ: $M \to N$ is said to be *regular* at $m \in M$ if the homomorphism ϕ_*: $H_m(M) \to H_{\phi(m)}(N)$ is injective. ϕ is called a *regular map* if it is regular at every $m \in M$. If we consider M, N as real analytic manifolds, we may consider the real map $\tilde{\phi}$ induced by ϕ. It is easy to check that ϕ is regular if and only if $\tilde{\phi}$ is regular.

The map ϕ also induces the two homomorphisms

$$\phi^* : H^*_{\phi(m)}(N) \to H^*_m(M)$$

and

$$\bar{\phi}^* : \bar{H}^*_{\phi(m)}(N) \to \bar{H}^*_m(M)$$

defined by $\phi^*(\partial f) = \partial(f \circ \phi)$ and $\bar{\phi}^*(\bar{\partial} f) = \bar{\partial}(f \circ \phi)$ respectively for $f \in \mathscr{F}_{\phi(m)}$.

By linearity, ϕ^* and $\bar{\phi}^*$ determine a homomorphism

$$\phi^* : E^{r,s}_{\phi(m)} \to E^{r,s}_m$$

of algebras of differential forms for every r, s.

Complex submanifolds, embeddings, and regular embeddings are defined following the lines of section 3.13.1.

We give here two interesting examples of complex manifolds.

i) Closed complex submanifolds regularly embedded in $\mathbf{P}^n(\mathbf{C})$, called *projective varieties.*

ii) Closed complex submanifolds regularly embedded in $\mathbf{C}^n$, called *Stein manifolds.*

We remark that a projective variety is a closed subset of a compact space and, therefore, is always compact.

A Stein manifold, which is not a single point, is never compact. In fact, if M is a Stein manifold and $m_1, m_2 \in M$, $m_1 \neq m_2$, there exists a coordinate function of $\mathbf{C}^n$ which takes different values at m_1, m_2. The restriction to M of that coordinate function is a global holomorphic function on M and it is non-constant. It follows by corollary 3.4 that M is not compact.

8.1 Problems

1 (1.1) If W is a finite dimensional complex vector space, there is an involution $J : W \to W$, called the complex structure of W, defined by $J(w) = iw$, with $J^2 = -1$

Let V be a finite dimensional real vector space. Show that:

i) $V \otimes \mathbf{C}$ is a complex vector space with complex structure defined by $J(v \otimes c) = v \otimes ic$, for $v \in V$, $c \in \mathbf{C}$,

ii) $\dim_{\mathbf{C}} V \otimes \mathbf{C} = \dim_{\mathbf{R}} V$,

iii) the conjugate of an element $v \otimes c$ is defined by $\overline{v \otimes c} = v \otimes \bar{c}$.

2 (1.1) Prove that the composition of two holomorphic functions is a holomorphic function.

3 (1.1) Show that the subset of $\mathbf{C}^n$ defined by the equations

$$z_j = \exp 2\pi i t_j;\ 0 \leq t_j \leq 1;\ j = 1, \ldots, n,$$

is an analytic manifold. Is it a complex manifold?

4 (2.1) Show that the real analytic structure of $\mathbf{P}^n(\mathbf{C})$ is different from that of $\mathbf{P}^{2n}(\mathbf{R})$.

5 (5.1) Prove that, if $w_1, \ldots, w_n$ are holomorphic functions at a point m of a complex manifold M and there is no functional relation between them, then $dw^1, \ldots, dw^n$ are linearly independent. Show that the converse is not necessarily true.

6 (6.1) Let M be an n-dimensional complex manifold and let $m \in M$. Consider the linear map $J: A^1_m \to A^1_m$ defined by $J(dz^j) = i\, d\bar{z}^j$ and $J(d\bar{z}^j) = -i\, dz^j$. Show that:

a) J is a real map and thus can be expressed as linear map $T^*_m \to T^*_m$ given by $dx^j = \sum_k \phi^j_k\, dx^k$,

b) the complex structure on M is completely determined by the vector field J,

c) the given complex structure on M implies the existence of a mixed tensor field on M, such that $\sum_k \phi^j_k\, \phi^k_l = -\delta^j_l$.

CHAPTER 5

Lie groups

1.1 Topological groups

WE SHALL CONSIDER a set G which has, simultaneously, a group structure and a topology. The topology on $G \times G$ will be the product space topology. The group operations $(g_1, g_2) \to g_1 g_2$ and $g \to g^{-1}$ are maps $G \times G \to G$ and $G \to G$ respectively. If G is a group, we shall denote by $\mathbf{e}$ the identity element of G.

1.2 DEFINITION *A set G is called a topological group if:*

i) *G is a Hausdorff topological space,*

ii) *G is a group,*

iii) *the group operations*

$$(g_1, g_2) \to g_1 g_2 \; (g_1, g_2 \in G)$$

and

$$g \to g^{-1} \; (g \in G)$$

are continuous.

Condition (iii) of definition 1.2 means that the topology of G must be compatible with the group structure. This condition (iii) is equivalent to:

iii′) *the map* $(g_1, g_2) \to g_1 g_2^{-1}$ *is continuous.*

In fact, if (iii′) holds, by putting $g_1 = \mathbf{e}$, it follows that $g_2 \to g_2^{-1}$ is continuous and, thus, $(g_1, g_2) \to g_1(g_2^{-1})^{-1} = g_1 g_2$ is also continuous. Conversely, if (iii) is valid, then $(g_1, g_2) \to (g_1, g_2^{-1})$ is a continuous map $G \times G \to G \times G$, and this implies that $(g_1, g_2) \to g_1 g_2^{-1}$ is a continuous map.

We remark that, if G is a topological group, the map $g \to g^{-1}$ is a homeomorphism of G onto itself.

Let H be any subgroup of a topological group G. It is easy to check that the topology induced on H by that of G is compatible with the group structure of H, and thus H is a topological group itself. Let G/H be the set of cosets of H in G. We remark that, in general, G/H does not have group structure. Let

$$p : G \to G/H$$

be the map defined by $p(g) = gH$ for $g \in G$.

1.3 LEMMA *Let H be a subgroup of a topological group G. Then G/H can be made into a topological space in such a way that the map $p: G \to G/H$ is continuous.*

Proof Define a set $U \subset G/H$ to be an open set if and only if $p^{-1}(U)$ is open in G. Since

$$\bigcup_{\alpha} p^{-1}(U_\alpha) = p^{-1}\left(\bigcup_{\alpha} U_\alpha\right),$$

and

$$\bigcap_{\alpha} p^{-1}(U_\alpha) = p^{-1}\left(\bigcap_{\alpha} U_\alpha\right),$$

the axioms for a topology on G/H are satisfied. Moreover, if U is open in G/H, then, by definition, $p^{-1}(U)$ is open in G. Thus p is a continuous map. ○

1.4 *Remark* It can be proved that the topology induced by G on G/H is the finest for which the map $p: G \to G/H$ is continuous.

1.5 LEMMA *The map $p: G \to G/H$ is an open map.*

Proof If U is an open set in G, then, for every $g \in G$, $gU = \{gu \mid u \in U\}$ is open. It follows that $HU = \bigcup_{h \in H} hU$ is open. On the other hand,

$$HU = p^{-1}(p(U))$$

which proves that $p(U)$ is open. ○

A subgroup H of a topological group G is called a *closed subgroup* if H is closed as a topological space.

In what follows we shall mostly consider closed subgroups.

1.6 *Remark* If H is a normal subgroup of G, then the topology induced by G on the factor group G/H is compatible with the group structure of G/H. In this case G/H satisfies conditions (ii) and (iii) of definition 1.2. However, G/H is not necessarily a topological group, because condition (i) may not be satisfied. It is easy to check that this condition is satisfied if H is a normal, closed subgroup of G.

Let X be a topological space and let G be a topological group. If

$$\phi: G \times X \to X$$

is a map, we shall denote by gx the point $\phi(g, x) \in X$.

1.7 DEFINITION *A topological group G is said to act on X if there exists a continuous map $\phi: G \times X \to X$ such that*

i) *for all* $x \in X$, $\mathbf{e}x = x$,
ii) *for all* $g, h \in G$, $x \in X$, $g(hx) = (gh)x$.

If G acts on X, G is also called a topological *transformation group* of X. More precisely, the action of G we considered is called left action; a right action is described similarly. The action of G depends, of course, on the given map ϕ.

If a topological group G acts on a topological space X, then for every fixed $g \in G$ the map $g: X \to X$ defined by $g(x) = gx$ is a homeomorphism of X onto itself since g has an inverse g^{-1} and g^{-1} is a continuous map. It follows that ϕ determines a homomorphism α of G into the group of homeomorphisms of X.

We say that G acts *effectively* on X if $gx = x$ for all $x \in X$ implies that $g = \mathbf{e}$.

If G acts effectively on X, then α is an isomorphism into and G may be considered as a group of homeomorphisms of X.

We say that G acts *transitively* on X if for every $x, y \in X$ there exists an element $g \in G$ such that $g(x) = y$.

Let G act transitively on X and let x be a fixed point of X. The set of elements of G leaving x fixed,

$$H = \{g \in G \mid g(x) = x\},$$

is a closed subgroup of G called the *isotropy* group of x.

We say that G acts *freely* on X if for all $x \in X$ we have $H = \mathbf{e}$, that is, if the only element of G having fixed points is the identity.

Let K be a closed subgroup of a topological group G. It is easy to see that G acts transitively on G/K when the action of G

$$\phi : G \times G/K \to G/K$$

is defined by

$$\phi(g, hK) = (gh)K$$

for $g, h \in G$.

If a topological group G acts transitively on a topological space X, then for every fixed $x \in X$ we may define a continuous map $x: G \to X$ by $x(g) = gx$.

Let H be the isotropy group of a fixed $x \in X$. For every $y \in X$, $x^{-1}(y)$ is an element of G/H. Hence we have a bijective map $\lambda: G/H \to X$ such that the following diagram

$$\begin{array}{ccc} G & \xrightarrow{p} & G/H \\ & {\scriptstyle x}\searrow \quad \swarrow{\scriptstyle \lambda} & \\ & X & \end{array}$$

commutes; that is, for every $g \in G$, $\lambda \circ p(g) = x(g)$. This map λ is clearly continuous but its inverse map, in general, is not continuous. However, in many cases λ is a homeomorphism. For instance, if G is compact, then G/H is compact and every injective map of a compact Hausdorff space onto a Hausdorff space is a homeomorphism.

2.1 Lie groups

Roughly speaking, a Lie group is a topological group which is also a manifold with some compatibility conditions.

2.2 DEFINITION *A topological group G is called a real Lie group if:*

i) *G is a differentiable manifold,*
ii) *the group operations* $(g_1, g_2) \to g_1 g_2$ *and* $g \to g^{-1}$ *are differentiable.*

2.3 DEFINITION *A topological group G is called a complex Lie group if:*

i) *G is a complex manifold,*
ii) *the group operations* $(g_1, g_2) \to g_1 g_2$ *and* $g \to g^{-1}$ *are holomorphic.*

A Lie group is not necessarily connected. If G is a Lie group, we shall denote by G^0 the connected component of G which contains the identity element $\mathbf{e}$ of G. G^0 is a closed subgroup of G. Every other connected component of G is homeomorphic to G^0. Therefore, if G is any Lie group, the (real or complex) dimension of G is well defined. It is the dimension of the manifold G^0.

It can be proved that a real Lie group is differentiably equivalent to a real analytic manifold with real analytic group operations. Moreover, every closed subgroup H of a Lie group G is a Lie group and the map $i: H \to G$ is analytic and regular. Then H is a closed submanifold of the manifold G.

Let G be a real Lie group and let M be a differentiable manifold. We say that G acts (differentiably) on M if G acts on M by a C^∞ map $\phi: G \times M \to M$. It follows that, for every $g \in G$, the map $g: M \to M$ is a diffeomorphism.

Let G be a complex Lie group and let M be a complex manifold. We say that G acts (holomorphically) on M if G acts on M by a holomorphic map $\phi: G \times M \to M$. It follows that, for every $g \in G$, the map $g: M \to M$ is a biholomorphic map.

If G is any topological group, every element $g \in G$ determines a surjective continuous map $\varrho_g: G \to G$ called *left translation*, defined for $h \in G$ by $\varrho_g(h) = gh$. A right translation ϱ_g^* is defined in a similar way. The inverse maps $\varrho_{g^{-1}}$ and $\varrho^*_{g^{-1}}$ are also continuous; therefore, left and right translations are homeomorphisms of G onto itself.

If G is a Lie group, left and right translations are real or complex analytic maps. It follows that every Lie group acts on itself by left and right translations.

Let H, G be Lie groups. A map $\sigma: H \to G$ is called a *homomorphism* if σ is a group homomorphism and a regular analytic map. $\sigma(H)$ is a subgroup of G and a submanifold of G.

The real line $\mathbf{R}$ is a Lie group under addition. If G is any Lie group and σ: $\mathbf{R} \to G$ is a homomorphism, $\sigma(\mathbf{R})$ is called a *one parameter subgroup* of G.

Let g, h be two elements of a Lie group G. There is a unique left translation $\varrho_{hg^{-1}}$ such that $\varrho_{hg^{-1}}(g) = h$. The map $\varrho_{hg^{-1}}: G \to G$ is an analytic homeomorphism which induces an isomorphism $(\varrho_{hg^{-1}})_*: T_g \to T_h$ of the tangent spaces to G at g and h.

2.4 DEFINITION *A vector field X on G is called left invariant if for every $g, h \in G$ we have $(\varrho_{hg^{-1}})_* (X_g) = X_h$.*

The definition of a right invariant vector field is similar.

2.5 THEOREM *Every one parameter subgroup of a Lie group G determines a left invariant vector field on G.*

Proof Let $x \in \mathbf{R}$. If σ: $\mathbf{R} \to G$ is a homomorphism, the map $x \to \sigma(x)$ may be considered in a generalized sense as a differentiable curve σ in G. If g is any element of G, then the map $x \to g\sigma(x)$ defines a differentiable curve $g\sigma$ in G as well. There is at least one of these curves through every point of G and it is not difficult to show that, if $m, g, h \in G$ and $m = g\sigma(x_1) = h\sigma(x_2)$, the curves $g\sigma$ and $h\sigma$ determine the same tangent vector in T_m. We can then define a vector field X on G by defining X_m to be the element of T_m determined by the curve $g\sigma$ if $m = g\sigma(x)$. The vector field X is left invariant, for, if ϱ_g is the left translation $\varrho_g(\sigma(x)) = m$, then we have $(\varrho_g)_* (X_{\sigma(x)}) = X_m$. ○

The converse of the above theorem is also true; that is, every left invariant vector field on G is determined by a one parameter subgroup of G, but we omit this proof.

3.1 Examples

We now give some examples of Lie groups, leaving most of the proofs to the reader.

3.2 *The general (real) linear group* $GL(n, \mathbf{R})$ It is the set of all $n \times n$ non-singular matrices with real entries. $GL(n, \mathbf{R})$ may be embedded in E^{m^2} in the following way: Let

$$(a_{ij}) = \begin{pmatrix} a_{11} & \cdots & a_{1n} \\ \cdot & \cdots & \cdot \\ \cdot & \cdots & \cdot \\ \cdot & \cdots & \cdot \\ a_{n1} & \cdots & a_{nn} \end{pmatrix}$$

be an element of $GL(n, \mathbf{R})$; define a map $\sigma\colon GL(n, \mathbf{R}) \to E^{n^2}$ by $\sigma(a_{ij}) = (a_{11}, \ldots, a_{1n}, a_{21}, \ldots, a_{nn})$. Then $GL(n, \mathbf{R})$ becomes an open subset of E^{n^2} and it is an open submanifold of E^{n^2} with the induced real analytic structure.

The group operation in $GL(n, \mathbf{R})$ is the product of matrices row by columns. Since the group operations are expressed by rational functions, they are real analytic. Thus $GL(n, \mathbf{R})$ is a real Lie group and $\dim_{\mathbf{R}} GL(n, \mathbf{R}) = n^2$. $GL(n, \mathbf{R})$ is not connected; it has two connected components consisting of matrices with positive or negative determinant respectively. The component of the identity of $GL(n, \mathbf{R})$ consists of matrices with positive determinant. $GL(n, \mathbf{R})$ is non-compact.

3.3 *The special (real) linear group* $SL(n, \mathbf{R})$ It is the subset of $GL(n, \mathbf{R})$ consisting of the matrices whose determinant is equal to 1. Since the determinant of the product of two matrices is the product of the determinants of these two matrices, $SL(n, \mathbf{R})$ is a subgroup of $GL(n, \mathbf{R})$. $SL(n, \mathbf{R})$ is a hypersurface of $GL(n, \mathbf{R})$. It is a connected real Lie group. It is non-compact and $\dim_{\mathbf{R}} SL(n, \mathbf{R}) = n^2 - 1$.

3.4 *The orthogonal group* $O(n)$ It is the subgroup of $GL(n, \mathbf{R})$ consisting of matrices $A \in GL(n, \mathbf{R})$ such that ${}^tA = A^{-1}$; that is, $O(n)$ consists of $n \times n$ orthogonal matrices. $O(n)$ is a closed subgroup of $GL(n, \mathbf{R})$. It is compact and non-connected since it contains matrices with determinant equal to 1 and matrices with determinant equal to -1. $O(n)$ is a real Lie group; its dimension is $\dfrac{n(n-1)}{2}$.

3.5 *The special orthogonal group* $SO(n)$ It is the connected component of the identity of $O(n)$. Therefore, we have

$$SO(n) = O(n) \cap SL(n, \mathbf{R}).$$

$SO(n)$ is a real compact connected Lie group; its dimension is, of course, the same as that of $O(n)$, that is, $\dfrac{n(n-1)}{2}$.

3.6 *The general (complex) linear group* $GL(n, \mathbf{C})$ It is the set of all $n \times n$ non-singular matrices with complex entries. With the same technique as in example 3.2, $GL(n, \mathbf{C})$ is an open complex submanifold of $\mathbf{C}^{n^2}$. The group operations in $GL(n, \mathbf{C})$ are complex analytic; therefore, $GL(n, \mathbf{C})$ is a complex Lie group, $\dim_{\mathbf{C}} GL(n, \mathbf{C}) = n^2$. It is a connected, non-compact Lie group.

Every element $g \in GL(n, \mathbf{C})$ is a non-singular linear transformation of $\mathbf{C}^n$. If $\{z_1, \ldots, z_n\}$ is a basis of $\mathbf{C}^n$, then $\{x_1, y_1, \ldots, x_n, y_n\}$ is a basis of $\mathbf{R}^{2n}$ and every non-singular linear transformation g of $GL(n, \mathbf{C})$ determines a linear transformation $\tilde{g} \in GL(2n, \mathbf{R})$. The map $g \to \tilde{g}$ makes $GL(n, \mathbf{C})$ into a subgroup of $GL(2n, \mathbf{R})$.

3.7 *The special (complex) linear group* $SL(n, \mathbf{C})$ It is the subgroup of $GL(n, \mathbf{C})$ of matrices whose determinant is equal to 1. It is a non-compact, connected, complex Lie group; $\dim_{\mathbf{C}} SL(n, \mathbf{C}) = n^2 - 1$. Clearly

$$SL(n, \mathbf{R}) = SL(n, \mathbf{C}) \cap GL(n, \mathbf{R}).$$

3.8 *The unitary group* $U(n)$ It is the subgroup of $GL(n, \mathbf{C})$ consisting of those matrices $A \in GL(n, \mathbf{C})$ such that ${}^tA = \bar{A}^{-1}$. If $A = (a_{ij})$, the condition ${}^tA = \bar{A}^{-1}$ is equivalent to the conditions

$$\sum_{i=1}^{n} a_{ij}\overline{a_{ik}} = \delta_{jk}.$$

It follows that $U(n)$ is not a complex submanifold of $GL(n, \mathbf{C})$. As a subspace of $GL(n, \mathbf{C})$, it can be embedded as a subgroup in $GL(2n, \mathbf{R})$. $U(n)$ is a connected, compact, real Lie group of dimension n^2.

3.9 *The special unitary group* $SU(n)$ It is defined as

$$SU(n) = SL(n, \mathbf{C}) \cap U(n).$$

4.1 Homogeneous spaces

If G is a topological group and H is a closed subgroup of G, the factor space G/H is a topological space and G acts transitively on G/H, as we have seen in section 1.1.

4.2 DEFINITION *Any topological space which may be obtained as the factor space of a topological group by a closed subgroup is called a homogeneous space.*

If G is a Lie group and H is a closed subgroup of G, then the factor space G/H is a homogeneous space.

Let G be a Lie group acting transitively on a topological space X. Let $x \in X$ and let H be the isotropy group of x. If the map $\lambda: G/H \to X$ is a homeomorphism, we may use it to transfer the differentiable structure of G/H to X. We remark that, since G acts transitively, if the map λ is a homeomorphism for one $x \in X$, then it is a homeomorphism for every $x \in X$.

If now G acts transitively on a differentiable manifold M, we may have two different differentiable structures on M. By using theorem 3.12.4, it is not difficult to prove that the differentiable structure of M and that induced on it by the map $m: G \to M$ coincide if and only if the map $m_*: T_g(G) \to T_{gm}(M)$ is surjective.

We shall now prove that the sphere $S^n \subset E^{n+1}$ is a homogeneous space relative to some subgroups of the general linear group.

4.3 THEOREM *For $n \geq 2$ the sphere S^{n-1} is homeomorphic to the factor space $O(n)/O(n-1)$.*

Proof The sphere S^{n-1} is the set of unit vectors in E^n. The orthogonal group $O(n)$ acts transitively on S^{n-1}. The isotropy group of the point $(0, \ldots, 0, 1) \in S^{n-1}$ consists of the matrices $A \in O(n)$ of the form

$$A = \begin{pmatrix} & & & 0 \\ & A' & & \vdots \\ & & & 0 \\ 0 & \cdots & 0 & 1 \end{pmatrix}.$$

This group is a closed subgroup H of $O(n)$ and, since the matrices A' are the elements of $O(n-1)$, then H is isomorphic to $O(n-1)$. The factor space $O(n)/O(n-1)$ is compact; therefore, the map

$$\lambda: O(n)/O(n-1) \to S^{n-1}$$

is a homeomorphism. ○

Actually, S^{n-1} and $O(n)/O(n-1)$ have the same differentiable structure since the map $s: O(n) \to S^{n-1}$ induces a surjective map $s_*: T_A(O(n)) \to T_{As}(S^{n-1})$.

4.4 THEOREM *For $n \geq 2$ the sphere S^{n-1} is homeomorphic to the factor space $SO(n)/SO(n-1)$.*

Proof It is enough to remark that if $(a_1, \ldots, a_n) \in S^{n-1}$ and $A \in O(n)$ is the matrix such that $A(0, \ldots, 0, 1) = (a_1, \ldots, a_n)$ with $\det A = -1$, we can always replace A by $\tilde{A} \in SO(n)$ where

$$\tilde{A} = A \begin{pmatrix} -1 & & & 0 \\ & \ddots & & \\ & & 1 & \\ & & & \ddots \\ 0 & & & 1 \end{pmatrix}. \quad \bigcirc$$

4.5 THEOREM *For $n \geq 2$ the sphere S^{2n-1} is homeomorphic to the factor space $U(n)/U(n-1)$.*

The proof is similar to that of theorem 4.3 since S^{2n-1} is the set of unit vectors in $\mathbf{C}^n$. ○

We shall now give some other examples of homogeneous spaces.

4.6 *Stiefel manifolds* Any ordered set of k independent vectors in E^n is called a k-frame. We shall denote by $V_{n,k}$ the set of all k-frames in E^n. Every k-frame is transformed into any other k-frame by an element of $GL(n, \mathbf{R})$. Therefore, although $V_{n,k}$ is not a topological space, we can say that $GL(n, \mathbf{R})$ acts transitively on $V_{n,k}$. If $v^k \in V_{n,k}$, we have a map

$$v^k : GL(n, \mathbf{R}) \to V_{n,k}$$

sending $(a_{ij}) \in GL(n, \mathbf{R})$ into $(a_{ij})\, v^k$. We define the open sets in $V_{n,k}$ to be the images under v^k of open sets of $GL(n, \mathbf{R})$. Because of the transitivity, this definition does not depend on the choice of the k-frame v^k. When $V_{n,k}$ is so topologized, $GL(n, \mathbf{R})$ acts continuously on $V_{n,k}$.

Let $H(n, k)$ be the isotropy group of a fixed k-frame $v^k \in V_{n,k}$; that is, every element of $H(n, k)$ leaves each vector of v^k fixed. The map

$$GL(n, \mathbf{R})/H(n, k) \to V_{n,k}$$

is a homeomorphism which can be used to transfer the differentiable structure of $GL(n, \mathbf{R})/H(n, k)$ to $V_{n,k}$. Endowed with this structure, $V_{n,k}$ is called the *Stiefel manifold of k-frames in E^n*.

We may restrict ourselves to the consideration of orthogonal k-frames in E^n, that is, k-frames such that every vector is of length one and every two vectors are orthogonal. The set of all orthogonal k-frames $S_{n,k}$ is a subset of $V_{n,k}$. In this case, the orthogonal group $O(n)$ acts transitively on $S_{n,k}$, and the isotropy group of a fixed orthogonal k-frame is isomorphic to $O(n-k)$. As in the previous case, the map

$$O(n)/O(n-k) \to S_{n,k}$$

makes $S_{n,k}$ into a differentiable manifold.

We remark that the Stiefel manifold $S_{n,1}$ is just the sphere S^{n-1}.

Consider the map

$$S_{n,k} \to S^{n-1}$$

given by $v^k = (v_1, \ldots, v_k) \to v_1$. This map is obtained by translating the k-frame v^k along its first vector to its end point on S^{n-1}. The remaining $k - 1$ vectors $(v_2, \ldots, v_k)$ form a $(k-1)$-frame of vectors tangent to S^{n-1}. Thus, the Stiefel manifold $S_{n,k}$ may be considered the manifold of all orthogonal $(k-1)$-frames tangent to S^{n-1}.

In particular, $S_{n,2}$ is the manifold of unit vectors tangent to S^{n-1}.

4.7 *Grassmann manifolds* A k-dimensional linear subspace of E^n is called a k-plane. We shall denote by $G_{n,k}$ the set of k-planes of E^n. As in example 4.6, although $G_{n,k}$ is not a topological space, the group $O(n)$ acts transitively on it. Let $E^k \in G_{n,k}$ and let $K(n, k)$ be the isotropy group of E^k. Any element of $K(n, k)$ leaves E^k fixed and it must, therefore, leave its orthogonal complement E^{n-k} fixed. It follows that $K(n, k)$ is isomorphic to $O(k) \times O(n-k)$. Hence, we have a homeomorphism

$$O(n)/O(k) \times O(n-k) \to G_{n,k},$$

and $G_{n,k}$ has the differentiable structure of that factor space. With this structure, $G_{n,k}$ is called the *Grassmann manifold of k-planes in* E^n.

We remark that $G_{n,1}$ is just the projective space $\mathbf{P}^{n-1}(\mathbf{R})$.

5.1 Problems

1 (1.1) Show that if G has discrete topology, this topology is always compatible with the group structure.

2 (1.1) Let $\mathbf{R}$ be the additive group of real numbers. Show that the topology defined by the metric $d(x, y) = |x - y|$ is compatible with the group structure.

3 (1.1) Let $\mathbf{R}$ be as before. Show that the topology of $\mathbf{R}$ defined by taking the half intervals $(a \leq \times \leq a + \delta)$, $\delta > 0$, as a basis neighborhoods at $a \in \mathbf{R}$ is not compatible with the group structure of $\mathbf{R}$.

4 (1.1) Let H be any subgroup of a topological group G. Show that if $\phi : G/H \to X$ is any map from G/H to a topological space X such that the map $\phi \circ p : G \to X$ is continuous, then ϕ is continuous.

5 (1.1) Use lemma 1.3 and the result of problem 4 to prove that the topology on G/H is uniquely determined and that it is the finest topology making the map $p : G \to G/H$ continuous.

6 (3.1) Compute the dimension of $O(n)$ and $U(n)$.

7 (4.1) Let G be a Lie group acting transitively on a differentiable manifold M. Prove that the differentiable structure induced on M by the map $m : G \to M$ coincides with the existing differentiable structure on M if and only if the map $m_* : T_g(G) \to T_{gm}(M)$ is surjective for every $g \in G$.

8 (4.1) Prove that the map $s_* : T_A(O(n)) \to T_{As}(S^{n-1})$ of theorem 4.3 is surjective.

CHAPTER 6

Fiber bundles

1.1 Fiber spaces

LET E, X BE two topological spaces and let $p: E \to X$ be a continuous map.

1.2 Definition *The triple (E, X, p) is called a fiber space, E is called the total space, X is called the base space, and p is called the projection. If $x \in X$, the set $E_x = p^{-1}(x)$ is called the fiber over x.*

Every fiber E_x has the topology which is induced on it by the inclusion in E. It is clear that $E = \bigcup_{x \in X} E_x$, every two distinct fibers E_x, E_y are disjoint if $x \neq y$, and, thus, every point of E is contained in one and only one fiber. Since we do not require the map p to be surjective, some of the fibers may be empty.

A fiber space (E, X, p) will be denoted simply by E when no confusion is possible. We shall also say that E is a fiber space over X.

Let $E = (E, X, p)$ be a fiber space and let Y be a subset of X. The *restriction* of E to Y is the fiber space $E \mid Y = (p^{-1}(Y), Y, p \mid p^{-1}(Y))$. We shall also write $E \mid Y = (E \mid Y, Y, p)$.

Let $E = (E, X, p)$ be a fiber space and let F be a subspace of E. The fiber space $F = (F, X, p \mid F)$ is called a *subfiber space* of E.

1.3 DEFINITION *Let E be a fiber space over X and let $Y \subset X$. A continuous map*

$$\sigma : Y \to E$$

such that, for every $x \in Y$, $p \circ \sigma(x) = x$, is called a cross-section of E over Y.

If $Y = X$, the cross-section σ is called a *global cross-section* of E. A cross-section σ of E over Y is a global cross-section of $E \mid Y$. By cross-section of E we shall always mean global cross-section.

Let E be a fiber space over X, let Y be a subset of X and let W be a subset of Y. If σ is a cross-section of E over Y, we shall denote by $\sigma| W$ the restriction of σ to W.

Let F be a subfiber space of a fiber space E over X and let σ be a cross-section of E. Then σ is a cross-section of F if and only if $\sigma(x) \in F$ for every $x \in X$.

Let $E = (E, X, p)$ be a fiber space. If $Y \subset X$, we shall denote by $\Gamma(Y, E)$ the set of all cross-sections of E over Y. If $W \subset Y$, the inclusion $W \to Y$ induces a restriction map:

$$\varrho_W^Y : \Gamma(Y, E) \to \Gamma(W, E).$$

It is easy to check that, if $V \subset W \subset Y$, the restriction maps satisfy the property

$$\varrho_V^W \circ \varrho_W^Y = \varrho_V^Y,$$

and, for every $Y \subset X$,

$$\varrho_Y^Y = \text{identity}.$$

Let $E = (E, X, p)$ and $F = (F, Y, q)$ be two fiber spaces. A pair (ϕ, ψ) of continuous maps $\phi\colon E \to F$, $\psi\colon X \to Y$ such that the following diagram

$$\begin{array}{ccc} E & \xrightarrow{\phi} & F \\ \downarrow{\scriptstyle p} & & \downarrow{\scriptstyle q} \\ X & \xrightarrow[\psi]{} & Y \end{array}$$

commutes is called a *morphism* or a *fiber map* and is denoted by $(\phi, \psi)\colon E \to F$.

Since for every $e \in E$ we have $q \circ \phi(e) = \psi \circ p(e)$, it follows that the pair $(e, p(e))$ is mapped by (ϕ, ψ) into $(\phi(e), \psi(p(e)))$. Hence, for every $x \in X$ we have $\phi(E_x) \subset F_{\psi(x)}$; that is, a fiber map maps fibers of E into fibers of F.

A particular case occurs when E, F are two fiber spaces over the same base space X and the map ψ is the identity. A fiber map $E \to F$ is given now by a single continuous map $\phi\colon E \to F$ such that the diagram

$$\begin{array}{ccc} E & \xrightarrow{\phi} & F \\ & {\scriptstyle p}\searrow \quad \swarrow{\scriptstyle q} & \\ & X & \end{array}$$

commutes. In this case, ϕ is called an *X-morphism.* For every $x \in X$ we now have $\phi(E_x) \subset F_x$.

Let $(\phi, \psi)\colon E \to F$ be a fiber map and let σ be a global cross-section of E. If the map $\psi\colon X \to Y$ is a homeomorphism, then,

$$\sigma' = \phi \circ \sigma \circ \psi^{-1} : Y \to F$$

is a global cross-section of F called the cross-section of F induced by (ϕ, ψ).

If E and F are two fiber spaces over the same base space X and $\phi\colon E \to F$

is a fiber map, to every cross-section σ of E over X there is always an induced cross-section σ' of F over X given by $\sigma' = \phi \circ \sigma$.

1.4 *Remark* Let $X = (X, X, i)$ be the fiber space defined by the identity map $i: X \to X$. Then, if $E = (E, X, p)$ is any fiber space over X, every cross-section σ of E can be defined as an X-morphism $\sigma: X \to E$.

Let $E = (E, X, p)$ and $F = (F, Y, q)$ be two fiber spaces. A fiber map $(\phi, \psi): E \to F$ is said to be an *isomorphism* if both ϕ and ψ are homeomorphisms. If (ϕ, ψ) is an isomorphism, then (ϕ^{-1}, ψ^{-1}) is a fiber map $F \to E$ and $(\phi \circ \phi^{-1}, \psi \circ \psi^{-1})$, $(\phi^{-1} \circ \phi, \psi^{-1} \circ \psi)$ are identities.

If E and F are two fiber spaces over the same base space X and $\phi: E \to F$ is a homeomorphism, then ϕ is called an *X-isomorphism.*

Two fiber spaces E, F over X are said to be *locally isomorphic* if to every point $x \in X$ there is a neighborhood U_x of x and an U_x-isomorphism between $E \mid U_x$ and $F \mid U_x$.

Let $p: E \to X$ be a fiber space over X; let Y be a topological space, and let $f: Y \to X$ be a continuous map. The *induced fiber space* f^*E is a fiber space $\pi: f^*E \to Y$ where the total space f^*E is the subspace of the product space $Y \times E$ consisting of those points (y, e), $y \in Y$, $e \in E$ such that $f(y) = p(e)$. The projection π is given by $\pi(y, e) = y$.

The restriction of a fiber space E over X to a subset $Y \subset X$ is a particular case of induced fiber space. In fact, if $i: Y \to X$ is the inclusion map, clearly i^*E is Y-isomorphic to $E \mid Y$.

The notion of a fiber space, as we defined it, is nothing more than the notion of a continuous function. We shall obtain a less trivial and rather useful object by requiring many additional structures.

2.1 Sheaves of sets

In this section we give an example of a particular fiber space which will often occur in the following chapters.

2.2 Definition *A fiber space (E, X, p) is called a sheaf of sets over X if every $e \in E$ has a neighborhood $U_e \subset E$ such that*

i) *the projection $p: U_e \to p(U_e)$ is a homeomorphism of U_e onto $p(U_e)$,*
ii) *$p(U_e)$ is a neighborhood of $p(e)$ in X.*

We remark that, if (E, X, p) is a sheaf of sets over X, the fiber E_x over $x \in X$ is of a very special nature. In fact, every point $e \in E_x$ has a neighborhood which does not contain any other point of E_x. It follows that E_x is a discrete space even though E may not have discrete topology.

2.3 THEOREM *Let (E, X, p) be a sheaf of sets over X, let Y be a subset of X and let σ be a cross-section of E over Y. To every $x \in Y$ there is a neighborhood $U \subset Y$ of x such that $\sigma \mid U = p^{-1} \mid U$.*

Proof Let $x \in Y$, let $e = \sigma(x)$, and let V be the neighborhood of e satisfying conditions (i) and (ii) of definition 2.2. Since σ is continuous, there exists a neighborhood U of x, $U \subset Y$, such that $\sigma(U) \subset V$. For every $x \in U$ we have $p \circ \sigma(x) = p \circ p^{-1}(x) = x$ and, therefore, $\sigma \mid U = p^{-1} \mid U$. ○

Later we shall see that if E is any fiber space over X the maps $Y \to \Gamma(Y, E)$, where Y is a subset of X, define a sheaf of sets over X.

3.1 Fiber bundles

Roughly speaking, a fiber bundle is a fiber space such that all the fibers are homeomorphic to a given space with a condition of a local triviality. The simplest example is a *product bundle* $(X \times F, X, \pi)$ where π is the natural projection $X \times F \to X$ defined by $\pi(x, e) = x$.

3.2 DEFINITION *A fiber space $E = (E, X, p)$ is called a fiber bundle if there exists a topological space F such that E is locally isomorphic to the product bundle $(X \times F, X, \pi)$.*

In other words, if E is a fiber bundle, to every $x \in X$ there are a neighborhood U of x in X and a homeomorphism

$$\phi : U \times F \to p^{-1}(U)$$

such that

$$p \circ \phi(y, e) = y$$

for every $y \in U$ and $e \in F$.

It follows from the definition that if the space E is not empty, the projection p is surjective and, for every $x \in X$, E_x is homeomorphic to F. If (F, E, X, p) is a fiber bundle, we shall denote it simply by E and say that E is a fiber bundle over X with fiber F and projection p.

Every fiber bundle which is isomorphic to a product bundle is called a *trivial bundle.*

Let E be a fiber bundle over X and let $\mathscr{U} = \{U_\alpha\}_{\alpha \in A}$ be an open covering of X such that $E \mid U_\alpha$ is a trivial bundle. The U_α-isomorphism ϕ_α:

$U_\alpha \times F \to E \mid U_\alpha$ will be called a *trivialization* of E over U_α. If $\alpha, \beta \in A$ and $U_{\alpha\beta} = U_\alpha \cap U_\beta \neq \emptyset$, there is a trivialization $\phi_{\alpha\beta}$ of E over $U_{\alpha\beta}$. Moreover, there is a homeomorphism $\theta_{\alpha\beta} : U_{\alpha\beta} \times F \to U_{\alpha\beta} \times F$ defined by $\theta_{\alpha\beta} = \bar{\phi}_\beta^{-1} \circ \bar{\phi}_\alpha$ where $\bar{\phi}_\alpha = \phi_\alpha \mid U_{\alpha\beta} \times F$ and $\bar{\phi}_\beta^{-1} = \phi_\beta^{-1} \, |E| \, U_{\alpha\beta}$.

The homeomorphism $\theta_{\alpha\beta}$ is clearly an $U_{\alpha\beta}$-isomorphism of the trivial bundle $U_{\alpha\beta} \times F$ onto itself and can be thought of as a family $\{\theta_x\}_{x \in U_{\alpha\beta}}$ of homeomorphisms of the fiber F onto itself.

Conversely, given the family $\{\theta_x\}_{x \in U_{\alpha\beta}}$ of homeomorphisms $\theta_x : F \to F$, one can define a continuous map

$$\theta_{\alpha\beta} : U_{\alpha\beta} \times F \to U_{\alpha\beta} \times F$$

by $\theta_{\alpha\beta}(x, e) = (x, \theta_x(e))$.

A familiar example of a fiber bundle is the Möbius band. It is a fiber bundle (F, E, X, p) where X is a circle obtained by identifying the end points of a line segment L; the fiber F is a line segment and the total space E is obtained from $L \times F$ by attaching the ends with a twist. The projection $p : E \to X$ is just the projection $L \times F \to L$.

Another example is the fiber bundle (S^1, S^3, S^2, p) defined as follows: Represent S^3 as the unit sphere in $\mathbf{C}^2$

$$S^3 = \{(z_1, z_2) \in \mathbf{C}^2 \mid z_1\bar{z}_1 + z_2\bar{z}_2 = 1\}$$

and S^2 as the complex projective line $p^1(\mathbf{C})$

$$S^2 = \{(w_1, w_2) \in \mathbf{C}^2\backslash\{0\} \mid (w_1, w_2) \sim (\lambda w_1, \lambda w_2), \lambda \neq 0\}.$$

Define a map

$$p : S^3 \to S^2$$

by $p(z_1, z_2) = (w_1, w_2)$ where $w_1 = z_1, w_2 = z_2$. Since $(w_1, w_2) \sim (w_1/(w_1\bar{w}_1 + w_2\bar{w}_2)^{1/2}, w_2/(w_1\bar{w}_1 + w_2\bar{w}_2)^{1/2})$, the map p, which is obviously continuous, is also surjective.

To prove the local triviality, let $U_1 = S^2\backslash\{(1, 0)\}$ and $U_2 = S^2\backslash\{(0, 1)\}$. Then $S^2 = U_1 \cup U_2$. Every point in U_1 is represented by $(w_1, 1)$ and every point in U_2 is represented by $(1, w_2)$. Let $S^1 = \{\lambda \in \mathbf{C} \mid |\lambda| = 1\}$. Define a map

$$\phi_1 : U_1 \times S^1 \to S^3$$

by

$$\phi_1((w_1, 1)\, \lambda) = (\lambda w_1/(w_1\bar{w}_1 + 1)^{1/2}, \quad \lambda/(w_1\bar{w}_1 + 1)^{1/2}).$$

It is easy to check that ϕ_1 is a homeomorphism of $U_1 \times S^1$ onto $p^{-1}(U_1)$ and that $p \circ \phi_1((w_1, 1), \lambda) = (w_1, 1)$ for every $(w_1, 1) \in U_1$ and $\lambda \in S^1$. Hence, ϕ_1 is a trivialization of S^3 over U_1.

A map $\phi_2 \colon U_2 \times S^1 \to S^3$ is defined similarly. Thus, S^3 is a fiber bundle over S^2 with fiber S^1.

The above fiber bundle was first considered by Hopf together with the bundles (S^3, S^7, S^4, p) and (S^7, S^{15}, S^8, p) constructed similarly by using quaternions and Cayley numbers respectively.

One more example of a fiber bundle will be given in the next section.

4.1 Tangent bundle

Let M be a differentiable manifold and let T_m be the tangent space to M at $m \in M$. Put

$$T(M) = \bigcup_{m \in M} T_m$$

and let $p \colon T(M) \to M$ be the natural projection which to every vector $t \in T_m$ associates the point $m \in M$.

If (U_α, u_α) is a local chart of M and $\dim M = n$, a map

$$\phi_\alpha : p^{-1}(U_\alpha) \to u_\alpha(U_\alpha) \times E^n$$

is defined as follows: Let $m \in U_\alpha$ and let $t \in T_m$ be represented by

$$\left(\frac{du_{\alpha 1}}{d\tau}, \ldots, \frac{du_{\alpha n}}{d\tau}\right)$$

in terms of the local coordinates $u_{\alpha 1}, \ldots, u_{\alpha n}$. Define

$$\phi_\alpha(m, t) = \left(u_{\alpha 1}, \ldots, u_{\alpha n}, \frac{du_{\alpha 1}}{d\tau}, \ldots, \frac{du_{\alpha n}}{d\tau}\right).$$

The map ϕ_α is injective and surjective and becomes a homeomorphism if we topologize $T(M)$ by defining open sets of $T(M)$ as inverse images under the maps ϕ_α of open sets in $u_\alpha(U_\alpha) \times E^n$.

It is clear that the map p is surjective and that, for every $m \in M$, $p^{-1}(m)$ is homeomorphic to E^n. Hence we have defined a fiber bundle $T(M) = (E^n, T(M), M, p)$ with fiber E^n. This fiber bundle is called the *tangent bundle* to the manifold M.

If now $\{U_\alpha\}$ is an open covering of M by coordinate neighborhoods, $\{(p^{-1}(U_\alpha), \phi_\alpha)\}$ is a basis of a differentiable structure for $T(M)$, which makes it into a differentiable manifold. In fact, if $u_{\alpha 1}, \ldots, u_{\alpha n}$ and $u_{\beta 1}, \ldots, u_{\beta n}$ are local coordinates of M in U_α and U_β respectively, ϕ_α and ϕ_β are C^∞-related since

$$\phi_\alpha \circ \phi_\beta^{-1} = \begin{cases} u_{\alpha j} = u_{\alpha j}(u_{\beta 1}, \ldots, u_{\beta n}) \\ \left(\dfrac{du_{\alpha j}}{d\tau}\right) = \left(\dfrac{\partial u_{\alpha j}}{\partial u_{\beta i}}\right)\left(\dfrac{du_{\beta i}}{d\tau}\right) \end{cases} \quad i, j = 1, \ldots, n$$

is a diffeomorphism $u_\beta(U_\alpha \cap U_\beta) \times E^n \to u_\alpha(U_\alpha \cap U_\beta) \times E^n$, that is, of two open sets of E^{2n}. The tangent bundle $T(M)$ is then a $2n$-dimensional differentiable manifold called the *manifold of tangent vectors to M*.

If M is a differentiable manifold, (E, M, p) is a fiber bundle over M, and U is a subset of M, we shall denote by $\Gamma_d(U, E)$ the set of differentiable cross-sections of E over U.

4.2 THEOREM *Let U be an open subset of a differentiable manifold M. Then* $\Gamma_d(U, T(M))$ *is the set of* C^∞ *contravariant vector fields over U.*

Proof If $\sigma: U \to T(M)$ is a cross-section, then for every $m \in U$ we have $p \circ \sigma(m) = m$, that is, the vector $\sigma(m) \in T_m$. Conversely, if $X = \{X_m\}$ is a vector field over U, the map which assigns the vector X_m to every $m \in U$ is clearly a cross-section of $T(M)$ over U. Moreover, the notion of differentiability is obviously the same. ○

If M is a complex manifold, the tangent bundle can always be defined with respect to the differentiable structure of M. In this case, a holomorphic tangent bundle can also be defined, and a result analogous to that of theorem 4.2 holds for holomorphic vector fields.

Of course, if M is any manifold, theorem 4.2 holds also with differentiable replaced by continuous.

It follows from theorem 4.2 that a vector field on a manifold M can be defined as a cross-section of $T(M)$ and this definition is totally independent of the local charts of M. Moreover, the problem of existence of continuous (or differentiable or holomorphic) vector fields on a manifold M is reduced to the problem of the existence of continuous (or differentiable or holomorphic) cross-sections of $T(M)$. For example, if $T(M)$ is a trivial bundle, there will be many cross-sections, since, in this case, cross-sections are just graphs of maps.

Given a manifold M, a *cotangent bundle* $T^*(M)$ can be defined in a similar way and analogous properties hold.

5.1 Coordinate bundles

With the notation of section 3.1, let (F, E, X, p) be a fiber bundle and let $\mathscr{U} = \{U_\alpha\}$ be an open covering of X such that $E \mid U_\alpha$ is a trivial bundle. We have already considered the trivializations

$$\phi_\alpha : U_\alpha \times F \to E \mid U_\alpha$$

and observed that, if $U_{\alpha\beta} = U_\alpha \cap U_\beta$, the homeomorphism $\theta_{\alpha\beta}$: $U_{\alpha\beta} \times F \to U_{\alpha\beta} \times F$ defined by $\theta_{\alpha\beta} = \phi_\beta^{-1} \circ \phi_\alpha$ can be considered a family of homeomorphisms of F onto itself, one for every $x \in U_{\alpha\beta}$.

Let $x \in U_\alpha \cap U_\beta$; we consider the composition

$$\{x\} \times F \xrightarrow{\phi_\alpha} p^{-1}(x) \xrightarrow{\phi_\beta^{-1}} \{x\} \times F$$

and, with a slight change of notation, put

$$g_{\alpha\beta}(x) = \phi_\beta^{-1} \circ \phi_\alpha.$$

5.2 DEFINITION *Let $E = (F, E, X, p)$ be a fiber bundle and let $\mathscr{U} = \{U_\alpha\}$ be an open covering of X such that, for every α, $E \backslash U_\alpha$ is trivial. Then E is called a coordinate bundle if*

i) *there exists a topological group G acting effectively on the fiber F,*

ii) *for every α, β and every $x \in U_\alpha \cap U_\beta$, $g_{\alpha\beta}(x) \in G$,*

iii) *for every α, β, $g_{\alpha\beta}(x)$ is a continuous function of x in $U_\alpha \cap U_\beta$.*

A coordinate bundle is denoted by (F, E, X, G, p); the group G is called the *structure group* of the bundle. The functions $g_{\alpha\beta}(x)$, defined on every non-empty intersection $U_\alpha \cap U_\beta$, are called the *transition functions* of the coordinate bundle. Since G acts effectively on F, the transition functions satisfy the property

5.3 $$g_{\alpha\beta}(x) \circ g_{\beta\gamma}(x) = g_{\alpha\gamma}(x), x \in U_\alpha \cap U_\beta \cap U_\gamma.$$

Putting $\alpha = \beta = \gamma$, it follows that

$$g_{\alpha\alpha}(x) = \text{identity}, \quad x \in U_\alpha,$$

and, for $\alpha = \gamma$, that

$$g_{\alpha\beta}(x) = g_{\beta\alpha}^{-1}(x), \quad x \in U_\alpha \cap U_\beta.$$

An example of a coordinate bundle is the tangent bundle $T(M)$ we have described in section 4.1. If M is a differentiable manifold of dimension n, the fiber of $T(M)$ is E^n and the structure group is the general linear group $GL(n, \mathbf{R})$. $T(M)$ is defined by any covering $\{U_\alpha\}$ of M by coordinate neighborhoods. If (U_α, u_α) and (U_β, u_β) are two local charts of M, the transition functions of the bundle are given on $U_\alpha \cap U_\beta$ by

$$g_{\alpha\beta}(m) = \left(\frac{\partial u_{\beta i}}{\partial u_{\alpha j}}\right)_m \in GL(n, \mathbf{R}).$$

The definition of a coordinate bundle depends on the open covering $\{U_\alpha\}$ of the base space and on the trivializations $\{\phi_\alpha\}$ relative to it.

Let $E = (F, E, X, G, p)$ be a coordinate bundle, let $\mathscr{U} = \{U_\alpha\}_{\alpha \in A}$ be the open covering of X, let $\phi_\alpha : U_\alpha \times F \to p^{-1}(U_\alpha)$ be the trivializations of E over U_α, and let $g_{\alpha\beta}(x)$ be the transition functions of the bundle.

Let $\mathscr{V} = \{V_j\}_{j \in J}$ be a refinement of the covering $\mathscr{U}$; that is, we have a map $\varrho : J \to A$ such that, for every $j \in J$, $V_j \subset U_{\varrho(j)}$. Then the open covering $\{V_j\}$ and the trivializations

$$\phi_{\varrho(j)} \mid V_j : V_j \times F \to p^{-1}(V_j)$$

define a coordinate bundle E' over X with transition functions $g_{\varrho(i)\varrho(j)}(x) \mid V_i \cap V_j$ which is different from E although $E' = (F, E, X, G, p)$.

Let $E_1 = (F, E_1, X, G, p_1)$ and $E_2 = (F, E_2, X, G, p_2)$ be two coordinate bundles with the same base space X, fiber F and structure group G. Let E_1 and E_2 be defined on the same open covering $\{U_\alpha\}$ of X with trivializations $\{\phi_\alpha\}$, $\{\psi_\alpha\}$ respectively. Let $\{g_{\alpha\beta}(x)\}$ be the transition functions of E_1 and $\{h_{\alpha\beta}(x)\}$ those of E_2.

5.4 DEFINITION *The coordinate bundles E_1, E_2 are said to be equivalent if there exists an X-isomorphism $\mu : E_1 \to E_2$ such that for every α, the homeomorphism $U_\alpha \times F \to U_\alpha \times F$ obtained by the composition*

$$U_\alpha \times F \xrightarrow{\phi_\alpha} p_1^{-1}(U_\alpha) \xrightarrow{\mu} p_2^{-1}(U_\alpha) \xrightarrow{\psi_\alpha^{-1}} U_\alpha \times F$$

is given by

$$(x, e) \to (x, g_\alpha(x)(e))$$

with $g_\alpha(x) \in G$ and $g_\alpha(x)$ continuous function of x in U_α.

If the two coordinate bundles E_1, E_2 are defined by using different open coverings $\{U_\alpha\}$, $\{V_i\}$ of X, we can consider a common refinement $\{W_j\}$ of $\{U_\alpha\}$ and $\{V_i\}$. We then say that E_1 and E_2 are equivalent if the induced bundles on $\{W_j\}$ are equivalent.

Let E_1, E_2 be two equivalent coordinate bundles and let $\mu: E_1 \to E_2$ be the X-isomorphism. For every $x \in X$ we have $\mu E_{1x} = E_{2x}$ and for every open set $U \subset X$ there is a one to one correspondence between $\Gamma(U, E_1)$ and $\Gamma(U, E_2)$. The restriction of μ to the fiber E_{1x} over x is an automorphism of the group G acting on the fiber F. Moreover, the following diagram

$$\begin{array}{ccccccc} \{x\} \times F & \xrightarrow{\phi_\alpha} & p_1^{-1}(x) & \xrightarrow{\mu} & p_2^{-1}(x) & \xrightarrow{\psi_\alpha^{-1}} & \{x\} \times F \\ \downarrow g_{\alpha\beta}(x) & & \| & & \| & & \downarrow h_{\alpha\beta}(x) \\ \{x\} \times F & \xrightarrow{\phi_\beta} & p_1^{-1}(x) & \xrightarrow{\mu} & p_2^{-1}(x) & \xrightarrow{\psi_\beta^{-1}} & \{x\} \times F \end{array}$$

shows that, for every α, β and $x \in U_\alpha \cap U_\beta$, the transition functions $g_{\alpha\beta}(x)$ and $h_{\alpha\beta}(x)$ are related by

$$h_{\alpha\beta}(x) = g_\beta(x) \circ g_{\alpha\beta}(x) \circ g_\alpha^{-1}(x).$$

5.5 THEOREM *If the structure group G of a coordinate bundle E consists of only one element, then E is a trivial bundle.*

Proof If $E = (F, E, X, G, p)$ and $G = \mathbf{e}$, then E is equivalent to the product bundle $X \times F$ since we can take $g_\alpha(x) = \mathbf{e}$. ○

From now on, by coordinate bundle we shall always mean an equivalence class of coordinate bundles. To represent this class we shall use any bundle in the class.

Let $E = (F, E, X, G, p)$ be a coordinate bundle and let $\mathscr{U} = \{U_\alpha\}$ be an open covering of X such that $E \mid U_\alpha$ is a trivial bundle. To E and $\mathscr{U}$ are associated the transition functions $g_{\alpha\beta}$ satisfying conditions 5.3. The next theorem will show a converse of this statement.

5.6 THEOREM *Let X, F, be two topological spaces, let G be a topological group acting effectively on F and let $\mathscr{U} = \{U_\alpha\}$ be an open covering of X. If for every α, β there are continuous maps $g_{\alpha\beta}: U_\alpha \cap U_\beta \to G$ such that*

5.3 $$g_{\alpha\beta}(x) \circ g_{\beta\gamma}(x), = g_{\alpha\gamma}(x), \; x \in U_\alpha \cap U_\beta \cap U_\gamma,$$

then there exists a unique coordinate bundle $E = (F, E, X, G, p)$ having the $g_{\alpha\beta}$ as transition functions.

Proof Let $\tilde{E}$ be the union of all disjoint products $U_\alpha \times F$. Identify two points $(x, e) \in U_\alpha \times F$ and $(y, f) \in U_\beta \times F$ if and only if

$$\begin{cases} x = y \\ f = g_{\alpha\beta}(x)(e). \end{cases}$$

Condition 5.3 shows that this is an equivalence relation. Define E to be the quotient space of $\tilde{E}$ by this equivalence relation with the induced topology. Denote by $[x, e]$ the equivalence class of (x, e). Define the projection $p: E \to X$ to be the natural projection $p\,[x, e] = x$. The map p is uniquely defined as it follows at once from the definition of equivalence and it is clearly continuous.

Now, for $x \in U_\alpha$ define $\phi_\alpha(x, e) = [x, e]$. Since $p\,[x, e] = x$, it follows that $p \circ \phi_\alpha(x, e) = x$. Thus, we have a map

$$\phi_\alpha : U_\alpha \times F \to p^{-1}(U_\alpha)$$

which is clearly continuous.

If now $[x, e] \in p^{-1}(U_\alpha)$, and $[x, e]$ is represented by (y, f) with $y \in U_\alpha \cap U_\beta$ and $f \in F$, then $y = x$ and $f = g_{\alpha\beta}(x)(e)$, which shows that $[x, e] = \phi_\alpha(x, g_{\alpha\beta}(x)(e))$. Thus ϕ_α is surjective.

If $[x, e] = [x', e']$, for $x \in U_\alpha$ we must have $x' = x$ and $e' = g_{\alpha\alpha}(x)(e) = e$, which proves that ϕ_α is injective.

It is easy to check that ϕ_α^{-1} is continuous; hence, ϕ_α is a homeomorphism which defines the trivialization of E over U_α.

Now, for $x \in U_\alpha \cap U_\beta$, x fixed, $\phi_\beta^{-1} \circ \phi_\alpha$ is a homeomorphism of F onto itself. Let $e' = \phi_\beta^{-1} \circ \phi_\alpha(e)$; then $\phi_\beta(e') = \phi_\alpha(e)$, which means that (x, e) is equivalent to (x, e'), that is, $e' = g_{\alpha\beta}(x)(e)$. Thus,

$$\phi_\beta^{-1} \circ \phi_\alpha = g_{\alpha\beta}(x).$$

This concludes the proof. ○

5.7 DEFINITION *A coordinate bundle $E = (F, E, X, G, p)$ is called a differentiable bundle if*

i) *X, E, F are differentiable manifolds,*

ii) *the projection p is a differentiable map,*

iii) *the structure group G is a real Lie group,*

iv) *the trivializations ϕ_α are differentiable homeomorphisms,*

v) *the transition functions $g_{\alpha\beta}$ are differentiable functions.*

A *holomorphic bundle* is defined likewise by replacing differentiable with complex analytic. If E is a holomorphic bundle over a complex manifold X and U is an open set contained in X, we shall denote by $\Gamma_0(U, E)$ the set of all holomorphic cross-sections of E over U.

5.8 THEOREM *Let E be a coordinate bundle over X. Then every point $x \in X$ has a neighborhood with a cross-section of E over it.*

Proof If $E = (F, E, X, G, p)$ is a coordinate bundle, the trivializations ϕ_α of E over U_α are continuous maps $\phi_\alpha : U_\alpha \times F \to E$. For every $e \in F$, let $j_e : U_\alpha \to U_\alpha \times \{e\}$ be defined by $j_e(x) = (x, e)$. j_e is a continuous map; then the composition

$$\phi_\alpha \circ j_e : U_\alpha \to E$$

is a cross-section of E over U_α. ○

The local cross-sections whose existence is proved in the above theorem are differentiable or holomorphic if E is a differentiable or holomorphic bundle respectively.

6.1 Principal bundles

The study of coordinate bundles is somewhat simplified by the introduction of the notion of a principal coordinate bundle.

6.2 DEFINITION *A coordinate bundle is called a principal coordinate bundle if the fiber coincides with the structure group acting on itself by left translations.*

An equivalence class of principal coordinate bundles is called a *principal bundle*. If $E = (G, E, X, G, p)$ is a principal bundle, the action of G on itself $G \times G \to G$ is defined by $(g, h) \to gh$.

If $E = (F, E, X, G, p)$ is any coordinate bundle, we define the *associated principal bundle* $\hat{E}$ to E to be the bundle $\hat{E} = (G, \hat{E}, X, G, \hat{p})$ where G acts on itself by left translations. The construction of the bundle $\hat{E}$ is as follows: Let the given coordinate bundle be defined with respect to the open covering $\{U_\alpha\}$ of X and let $g_{\alpha\beta}$ be the transition functions of E. Then $\hat{E}$ is the bundle with base space X, fiber G, structure group G, and transition functions $g_{\alpha\beta}$, whose existence and uniqueness was proved in theorem 5.6.

6.3 THEOREM *Two coordinate bundles with the same base space, fiber and structure group are equivalent if and only if their associated principal bundles are equivalent.*

Proof Let E_1, E_2 be two coordinate bundles over X and let $\{U_\alpha\}$ be an open covering of X such that $E_1 \mid U_\alpha$ and $E_2 \mid U_\alpha$ are trivial bundles.

With respect to the covering $\{U_\alpha\}$, E_1 and its associated principal bundle $\hat{E}_1$ have the same transition functions $g_{\alpha\beta}(x)$. Similarly, E_2 and $\hat{E}_2$ have the same transition functions $h_{\alpha\beta}(x)$.

The proof of the theorem follows now from the fact that equivalence of coordinate bundles is a property involving only transition functions. In fact, as we remarked in section 5.1, E_1 is equivalent to E_2 if and only if there exist continuous functions g_α such that

$$h_{\alpha\beta}(x) = g_\beta(x) \circ g_{\alpha\beta}(x) \circ g_\alpha^{-1}(x). \quad \bigcirc$$

Let G be a topological group acting transitively on a topological space X. Let H be the isotropy group of $x \in X$. Let $p: G \to G/H$ be defined by $p(g) = gH$. We shall prove that under certain hypotheses G is a principal bundle over G/H with fiber H and structure group H.

6.4 THEOREM *Let G be a topological group and let H be a closed subgroup of G. If every point $\alpha \in G/H$ has a neighborhood U over which there is a cross-section of $p: G \to G/H$, then G is a principal bundle over G/H with fiber H and structure group H acting on itself by left translations.*

Proof Let $\alpha \in G/H$. By hypothesis there are a neighborhood U_α of α in G/H and a cross-section $\sigma_\alpha : U_\alpha \to G$. The map

$$\phi_\alpha : U_\alpha \times H \to G$$

defined, for $x \in U_\alpha$ and $h \in H$, by $\phi_\alpha(x, h) = \sigma_\alpha(x)\, h$ is a homeomorphism. Since σ_α is a cross-section, we have $p \circ \sigma_\alpha(x) = x$ and $p \circ \phi_\alpha(x, h) = x$. Thus, ϕ_α maps $U_\alpha \times H$ onto $p^{-1}(U_\alpha)$.

The family $\{U_\alpha\}_{\alpha \in G/H}$ is an open covering of G/H; hence, the homeomorphisms

$$\phi_\alpha : U_\alpha \times H \to p^{-1}(U_\alpha)$$

provide the trivializations of G over U_α.

If now $x \in U_\alpha \cap U_\beta$ and $h \in H$, there exists $f \in H$ such that $\sigma_\alpha(x)\, h = \sigma_\beta(x) f$. If we define

$$g_{\alpha\beta}(x) = hf^{-1}$$

we have

$$\sigma_\alpha(x)\, g_{\alpha\beta}(x) = \sigma_\beta(x),$$

and then for every $h \in H$ we have

$$\phi_\alpha(x, g_{\alpha\beta}(x)(h)) = \phi_\beta(x, h).$$

It is not difficult to prove that, if we start with a different cross-section, we obtain an equivalent principal bundle.

6.5 *Remark* There is a more general version of the previous theorem which we will not prove. If K is a closed subgroup of H and H is a closed subgroup of a topological group G, then, under the same hypothesis of theorem 6.4, there is a coordinate bundle $p: G/K \to G/H$ with fiber H/K. The structure group of the bundle is H/K_0 where K_0 is the largest subgroup of K which is invariant in H. The group H/K_0 acts on H/K by left translations.

The problem of existence of local cross-sections, which is essential in theorem 6.4, is a very difficult one when G and H are any two Lie groups. Chevalley [6] has proved that local cross-sections always exist if G is a Lie group and H is a closed subgroup of G. We shall only consider this situation in order to produce some examples which arise from the homogeneous spaces we considered in 5.4.1.

Stiefel manifolds provide our first example. The Stiefel manifold $S_{n,k}$ is homeomorphic to $O(n)/O(n-k)$. We have $S_{n,n} = O(n)$ and $S_{n,1} = S^{n-1}$.

Since $O(n-k-1) \subset O(n-k)$, the inclusion of cosets gives rise to a chain of Stiefel manifolds and maps

$$O(n) = S_{n,n} \to S_{n,n-1} \to \cdots \to S_{n,2} \to S_{n,1} = S^{n-1}.$$

By applying the theorem of remark 6.5 to the chain of inclusions $O(n-k-1) \subset O(n-k) \subset O(n)$, we get that $S_{n,k+1}$ is a coordinate bundle over $S_{n,k}$ with fiber S^{n-k-1} and structure group $O(n-k)$.

A coordinate bundle with fiber S^n and structure group $O(n+1)$ is called an *n-sphere bundle*. We have then a chain of sphere bundles connecting S^{n-1} to $O(n)$.

In order to show that $S_{n,k+1} \to S_{n,k}$ is a coordinate bundle with structure group $O(n-k)$, we have used the fact that the largest subgroup of $O(n-k-1)$ which is invariant in $O(n-k)$ consists only of the identity element.

A second example is the orthogonal group. $O(n)$ is a principal bundle over S^{n-1} with fiber $O(n-1)$ and structure group $O(n-1)$ as it follows at once from theorem 6.4 applied to the inclusion $O(n-1) \subset O(n)$.

We remark that, for every k, the Stiefel manifold $S_{n,k}$ is a coordinate bundle over S^{n-1}. It can be proved that for every k the principal bundle associated to $S_{n,k} \to S^{n-1}$ is $O(n) \to S^{n-1}$. It can also be proved that these bundles are trivial for $n = 2, 4, 8$.

7.1 Reduction of the structure group

Let E be a coordinate bundle over X with fiber F and structure group G. It may happen that the equivalence class of E contains a coordinate bundle having as structure group a subgroup H of G. In this case, there is an open covering $\{U_\alpha\}$ of X such that for every α, β and every $x \in U_\alpha \cap U_\beta$ the transition functions $g_{\alpha\beta}(x)$ are elements of H. We then say that *the structure group of E can be reduced to H.*

Let M be a differentiable manifold of dimension n and let $\{U_\alpha\}$ be an open covering of M by coordinate neighborhoods. The cotangent bundle $T^*(M)$ has structure group $GL(n, \mathbf{R})$ and trivializations ϕ_α: $U_\alpha \times E^n \to p^{-1}(U_\alpha)$.

7.2 DEFINITION *A Riemannian metric on M is the assignment on every local chart* (U_α, u_α) *of a symmetric quadratic form on* $T^*(M) \mid U_\alpha$

$$ds_\alpha^2 = \sum_{i,j=1}^{n} g_{ij} du_\alpha^i du_\alpha^j$$

such that

i) *for every* i, j, g_{ij} *is a differentiable function on* U_α,

ii) *the matrix* $(g_{ij})_m$ *is positive definite at every* $m \in U_\alpha$,

iii) *for every* α, β, $ds_\alpha^2 = ds_\beta^2$.

Condition (i) says that the entries of the matrix (g_{ij}) are differentiable functions of the local coordinates, condition (ii) that $ds_\alpha^2 > 0$ and condition (iii) that in $U_\alpha \cap U_\beta$ we must have

$$g_{hk}(u_\beta) = \sum_{i,j=1}^{n} g_{ij}(u_\alpha) \frac{\partial u_{\alpha i}}{\partial u_{\beta h}} \frac{\partial u_{\alpha j}}{\partial u_{\beta k}}.$$

7.3 THEOREM *Every differentiable manifold can be given a Riemannian metric.*

Proof Let $\{U_\alpha\}$ be an open covering of M by coordinate neighborhoods and let $\{\phi_\alpha\}$ be a C^∞ partition of unity subordinate to the covering $\{U_\alpha\}$. Consider on (U_α, u_α) the quadratic form

$$ds_\alpha^2 = \sum_{i=1}^{n} (du_\alpha^i)^2;$$

then

$$ds^2 = \sum_\alpha \phi_\alpha ds_\alpha^2$$

is a Riemannian metric on M.

Let M, N be two differentiable manifolds of dimensions m and n and let $\phi: M \to N$ be a regular map. It is easy to show that every Riemannian metric on N induces a Riemannian metric on M. In fact, let (V, v) be a local chart of N and let

$$ds^2 = \sum_{i,j=1}^{n} g_{ij} dv^i dv^j$$

be the expression of the metric on V. If, for $u \in M$, we put

$$\tilde{g}_{hk}(u) = \sum_{i,j=1}^{n} g_{ij}(\phi(u)) \frac{\partial v_i}{\partial u_h} \frac{\partial v_j}{\partial u_k},$$

then

$$ds^2 = \sum_{h,k=1}^{m} \tilde{g}_{hk} du^h du^k$$

is the expression, in the local coordinates on a neighborhood of u, of a Riemannian metric on M.

7.4 THEOREM *The assignment of a Riemannian metric on a differentiable manifold M is equivalent to the reduction of the structure group of $T^*(M)$ to the orthogonal group.*

Proof In a local chart (U_α, u_α) of M, a Riemannian metric is given by a matrix (g_{ij}) which is symmetric and positive definite at every point $m \in U_\alpha$. Thus, the matrix $(g_{ij})_m$ can be diagonalized. In fact, there exist matrices A such that

$${}^tA(g_{ij})_m\, A = I$$

where I is the identity matrix.

The matrix A is a differentiable function of $m \in U_\alpha$. If we multiply A^{-1} by $(du_\alpha^1, \ldots, du_\alpha^n)$, we get an n-tuple of linearly independent differential forms $(\omega_\alpha^1, \ldots, \omega_\alpha^n)$ of degree 1 defined on U_α such that

$$ds_\alpha^2 = \sum_{i,j=1}^{n} g_{ij} du_\alpha^i du_\alpha^j = \sum_{i=1}^{n} (\omega_\alpha^i)^2.$$

We can take $\{\omega_\alpha^1, \ldots, \omega_\alpha^n\}$ as a basis for the cotangent space to M on U_α. Since on $U_\alpha \cap U_\beta$ we have

$$ds_\alpha^2 = ds_\beta^2 = \sum_{i=1}^{n} (\omega_\alpha^i)^2 = \sum_{i=1}^{n} (\omega_\beta^i)^2,$$

it follows that $\{\omega_\alpha^1, \ldots, \omega_\alpha^n\}$ is transformed into $\{\omega_\beta^1, \ldots, \omega_\beta^n\}$ by an orthogonal matrix; that is, in the local charts $(U_\alpha, \omega_\alpha)$ the structure group of $T^*(M)$ is $O(n)$.

Conversely, let the structure group of $T^*(M)$ be reduced to $O(n)$. Let

$$\phi_\alpha : U_\alpha \times E^n \to p^{-1}(U_\alpha)$$

be the trivialization of $T^*(M)$ over U_α and let $\omega_\alpha^1, \ldots, \omega_\alpha^n$ be the n-tuple of differential forms induced on $p^{-1}(U_\alpha)$ by the basis on $U_\alpha \times E^n$. Put

$$ds_\alpha^2 = \sum_{i=1}^{n} (\omega_\alpha^i)^2.$$

Conditions (i) and (ii) of definition 7.2 are clearly satisfied. Since the structure group of $T^*(M)$ is $O(n)$, on $U_\alpha \cap U_\beta$ the n-tuples $(\omega_\alpha^1, \ldots, \omega_\alpha^n)$ and $(\omega_\beta^1, \ldots, \omega_\beta^n)$ are related by an orthogonal matrix and thus we have $ds_\alpha^2 = ds_\beta^2$, which proves that condition (iii) is also satisfied. Hence, the ds_α^2 determine a Riemannian metric on M. ○

It follows from the previous theorem that the structure group of the tangent bundle $T(M)$ can also be reduced to $O(n)$.

It is also easy to verify that, if M is orientable, the structure group of $T(M)$ can be reduced to $SO\,(n)$.

If M is a complex manifold, the structure groups of the complex tangent and cotangent bundles $H(M)$, $H^*(M)$, considered as complex-differentiable bundles, can be reduced to the unitary group $U(n)$. However, the reduction to $U(n)$ does not make any sense if we consider the holomorphic structure of these bundles, since $U(n)$ is not a complex Lie group.

8.1 Vector bundles

If M is a differentiable manifold, the tangent bundle $T(M)$ and the cotangent bundle $T^*(M)$ have the property that the fiber is a vector space and the structure group is the general linear group. We shall now consider this special class of bundles.

8.2 DEFINITION *A coordinate bundle is called a vector bundle if the fiber is a finite dimensional vector space and the structure group is the general linear group.*

We shall allow the structure group of a vector bundle to be a subgroup of the general linear group in view of the fact that the group of a bundle is sometimes reducible to a subgroup.

Infinite dimensional vector bundles can also be considered. Their definition is obvious but their properties are less elementary.

We are interested in the following different types of vector bundles:

8.3 *n-dimensional real vector bundles* The base space is a topological space X, the fiber is the real vector space E^n, the structure group is $GL(n, \mathbf{R})$ acting on E^n as a topological group of automorphisms, and the trivializations ϕ_α and the transition functions $g_{\alpha\beta}$ are continuous functions.

8.4 *n-dimensional real differentiable vector bundles* The base space is a differentiable manifold M, the fiber is the real vector space E^n, the structure group is $GL(n, \mathbf{R})$ considered as a real Lie group, and the trivializations ϕ_α and the transition functions $g_{\alpha\beta}$ are differentiable functions.

8.5 *n-dimensional complex-differentiable vector bundles* The base space is a differentiable manifold M, the fiber is the complex vector space $\mathbf{C}^n$, the structure group is $GL(n, \mathbf{C})$ considered as a real Lie group, and the trivializations ϕ_α and the transition functions $g_{\alpha\beta}$ are differentiable functions.

8.6 *n-dimensional holomorphic vector bundles* The base space is a complex manifold M, the fiber is the complex vector space $\mathbf{C}^n$, the structure group is $GL(n, \mathbf{C})$ considered as a complex Lie group, and the trivializations ϕ_α and the transition functions $g_{\alpha\beta}$ are holomorphic functions.

Every n-dimensional holomorphic vector bundle can always be considered as an n-dimensional complex differentiable vector bundle and as a $2n$-dimensional real differentiable vector bundle.

A 1-dimensional vector bundle is also called a *line-bundle.*

For every base space X there exists a unique zero-dimensional vector bundle.

If $p:E \to X$ is a vector bundle with fiber E^n, for every $x \in X$ the fiber E_x over x has a natural vector space structure. In fact, if $\{U_\alpha\}$ is an open covering of X such that $E \mid U_\alpha$ is a trivial bundle, we can use the trivializations

$$\phi_\alpha : U_\alpha \times E^n \to p^{-1}(U_\alpha)$$

to transfer the vector space structure of E^n to $E_x = p^{-1}(x)$. This is easily done by defining

$$a_1\phi_\alpha(x, v_1) + a_2\phi_\alpha(x, v_2) = \phi_\alpha(x, a_1v_1 + a_2v_2)$$

for $x \in U_\alpha$; $v_1, v_2 \in E^n$; $a_1, a_2 \in \mathbf{R}$. It is a trivial matter to show that the definition is independent of α.

For every $x \in U_\alpha$ we can also define the *zero cross-section* of E over U_α to be the map $x \to \phi_\alpha(x, 0)$.

It follows that, given two vector bundles E_1 and E_2 over the same base space X, for every $x \in X$ we may consider the direct sum, tensor product, exterior product and any other operation of the two vector spaces E_{1x}, E_{2x}. We shall show that every operation performed on the single fibers gives rise to a new vector bundle over X.

Let $p_1 : E_1 \to X$ be a vector bundle with fiber F_1 of dimension n_1 and let $p_2 : E_2 \to X$ be a vector bundle with fiber F_2 of dimension n_2. Since both E_1 and E_2 have the same base space, we can assume that they are defined over the same open covering $\{U_\alpha\}$ of X. With respect to that covering, let $\overset{1}{g}_{\alpha\beta}$ and $\overset{2}{g}_{\alpha\beta}$ be the transition functions of E_1, E_2 respectively. We shall describe a few examples of operations.

8.7 *The direct sum or Whitney sum* $E = E_1 \oplus E_2$ is a vector bundle over X with fiber $F = F_1 \oplus F_2$ of dimension $n_1 + n_2$. The transition functions $g_{\alpha\beta}$ of the direct sum E with respect to the covering $\{U_\alpha\}$ must operate on the first summand as $\overset{1}{g}_{\alpha\beta}$ and on the second summand as $\overset{2}{g}_{\alpha\beta}$. Thus, they are given, for $x \in U_\alpha \cap U_\beta$, by

$$g_{\alpha\beta}(x) = \begin{pmatrix} \overset{1}{g}_{\alpha\beta}(x) & 0 \\ 0 & \overset{2}{g}_{\alpha\beta}(x) \end{pmatrix}.$$

8.8 *The tensor product* $E = E_1 \otimes E_2$ is a vector bundle over X with fiber $F = F_1 \otimes F_2$ of dimension $n_1 n_2$.

Let $\{v_i\}_{i=1,\ldots,n_1}$ and $\{w_j\}_{j=1,\ldots,n_2}$ be bases of F_1, F_2. A basis of F is given by $\{v_i \otimes w_j\}$, $i = 1, \ldots, n_1$; $j = 1, \ldots, n_2$. The automorphism $\overset{1}{g}_{\alpha\beta} : F_1 \to F_1$ maps the basis $\{v_i\}$ into a new basis $\{v'_i\}$ and the automorphism $\overset{2}{g}_{\alpha\beta} : F_2 \to F_2$ maps the basis $\{w_j\}$ into a new basis $\{w'_j\}$. One has

$$(v'_i)^k = \sum_{h=1}^{n_1} (\overset{1}{g}_{\alpha\beta})_{hk}\, v_i^h$$

and

$$(w'_j)^r = \sum_{s=1}^{n_2} (\overset{2}{g}_{\alpha\beta})_{sr}\, w_j^s.$$

The induced automorphism of $F = F_1 \otimes F_2$ must take the basis $\{v_i \otimes w_j\}$ into the basis $\{v'_i \otimes w'_j\}$. Therefore, it is represented by a matrix such that

$$\textbf{8.9} \qquad (v'_i \otimes w'_j)^{k,r} = \sum_{h=1}^{n_1} \sum_{s=1}^{n_2} (\overset{1}{g}_{\alpha\beta})_{hk} (\overset{2}{g}_{\alpha\beta})_{sr}\, v_i^h w_j^s.$$

Thus, the transition functions of the tensor product bundle $E = E_1 \otimes E_2$ are given by the tensor product of the matrices $\overset{1}{g}_{\alpha\beta}$, $\overset{2}{g}_{\alpha\beta}$; that is, for $x \in U_\alpha \cap U_\beta$

$$g_{\alpha\beta}(x) = (\overset{1}{g}_{\alpha\beta}(x)) \otimes (\overset{2}{g}_{\alpha\beta}(x)).$$

8.10 *The exterior product* $E = E_1 \wedge E_1$ is a vector bundle over X with fiber $F = F_1 \wedge F_1$ of dimension $\binom{n_1}{2}$.

If $\{v_i\}$ is a basis of F_1 as before, then $\{v_i \wedge v_j\}_{i<j}$ is a basis of F. Since $v_i \wedge v_j = -v_j \wedge v_i$, the sum in 8.9 may be reduced to $1 \leq h < s \leq n_1$. If we denote by $\overset{1}{g}_{\alpha\beta} \wedge \overset{1}{g}_{\alpha\beta}$ the matrix so obtained, the transition functions of the vector bundle $E = E_1 \wedge E_1$ are given, for $x \in U_\alpha \cap U_\beta$, by

$$g_{\alpha\beta}(x) = (\overset{1}{g}_{\alpha\beta}(x)) \wedge (\overset{1}{g}_{\alpha\beta}(x)).$$

The tangent and cotangent bundles to a differentiable manifold provide familiar examples of vector bundles. We shall also consider the following vector bundles related to a manifold M.

With base space an n-dimensional differentiable manifold M:
$A^p(M) = \wedge^p T^*(M)$ is a real differentiable vector bundle of dimension $\binom{n}{p}$. Elements of $\Gamma_d(U, A^p(M))$ are real valued differential forms of degree p defined on an open set $U \subset M$.

With base space an n-dimensional complex manifold M:
$E^{p,0}(M) = \wedge^p H^*(M)$ is a holomorphic vector bundle of dimension $\binom{n}{p}$. Elements of $\Gamma_0(U, E^{p,0}(M))$ are holomorphic differential forms of type $(p, 0)$ defined on $U \subset M$.

$E^{0,q}(M) = \wedge^q \bar{H}^*(M)$ is a complex differentiable vector bundle of dimension $\binom{n}{q}$. Elements $\Gamma_d(U, E^{0,q}(M))$ are antiholomorphic differential forms of type $(0, q)$ defined on $U \subset M$.

$E^{p,q}(M) = (\wedge^p H^*(M)) \wedge (\wedge^q \bar{H}^*(M))$ is a complex differentiable vector bundle of dimension $\binom{n}{p}\binom{n}{q}$. Elements of $\Gamma_d(U, E^{p,q}(M))$ are complex valued differential forms of type (p, q) defined on $U \subset M$.

$E^h(M) = \bigoplus_{p+q=h} E^{p,q}(M)$ is a complex differentiable vector bundle of dimension $\binom{2n}{h}$. Elements of $\Gamma_d(U, E^h(M))$ are complex valued differential forms of degree h defined on $U \subset M$.

9.1 Vector bundle homomorphisms

Let $p_1: E_1 \to X_1$ be an m-dimensional vector bundle and let $p_2: E_2 \to X_2$ be an n-dimensional vector bundle.

9.2 DEFINITION *A fiber map $(\phi, \psi): E_1 \to E_2$ is called a vector bundle homomorphism or simply a homomorphism if the restriction of ϕ to every fiber E_{1x} is a vector space homomorphism.*

If $m = n$ and $\phi \mid E_{1x}$ is of rank n for every $x \in X$, the homomorphism (ϕ, ψ) is called a *bundle isomorphism.*

If E_1, E_2 are vector bundles over the same base space X, an X-morphism $\phi: E_1 \to E_2$ is called an *X-homomorphism* if ϕ is a homomorphism.

An X-homomorphism which is also a bundle isomorphism is called a *vector bundle isomorphism* or simply an isomorphism.

Two isomorphic vector bundles are equivalent; that is, a vector bundle is defined up to an isomorphism.

Let $\phi: E_1 \to E_2$ be an X-homomorphism. Let $\sigma^0: X \to E_2$ be the zero cross-section of E_2. The set $\phi^{-1}(\sigma^0(X))$ is called the *kernel* of ϕ. In general, the kernel of ϕ is not a vector bundle because it may not satisfy the local triviality property (see problem 8).

If ϕ is surjective, then rank $\phi \mid E_{1x} = n$ for all $x \in X$. It follows that the kernel of ϕ is the total space of an $(m - n)$-dimensional vector bundle over X denoted by Ker ϕ.

If $\phi: E_1 \to E_2$ is any X-homomorphism, we identify $e_2, e_2' \in E_2$ if $p_2(e_2) = p_2(e_2')$ and $e_2 - e_2' \in \phi(E_1)$. This is clearly an equivalence relation. The quotient space of the total space E_2 by this equivalence relation is called the *cokernel* of ϕ. It is easy to show that, if ϕ is injective, the cokernel of ϕ is the total space of a vector bundle over X, denoted by Coker ϕ, with fiber of dimension $n - m$.

Let E_1, E_2, E_3 be vector bundles over X, let $\phi: E_1 \to E_2$ be an injective X-homomorphism and $\eta: E_2 \to E_3$ be a surjective X-homomorphism. Then the sequence

$$0 \to E_1 \xrightarrow{\phi} E_2 \xrightarrow{\eta} E_3 \longrightarrow 0$$

of vector bundles and X-homomorphisms is exact if $\phi(E_1) = \text{Ker}\, \eta$.

If $p_1: E_1 \to X$ and $p_2: E_2 \to X$ are two vector bundles over X, the total space E of the vector bundle $E = E_1 \oplus E_2$ consists of points (e_1, e_2) such that $p_1(e_1) = p_2(e_2)$. If we denote by $\sigma_1^\circ: X \to E_1$ and $\sigma_2^\circ: X \to E_2$ the zero cross-sections of E_1 and E_2 respectively, the direct sum diagram

$$E_1 \underset{j_1}{\overset{\pi_1}{\leftrightarrows}} E_1 \oplus E_2 \underset{j_2}{\overset{\pi_2}{\rightleftarrows}} E_2$$

consists of projections π_1, π_2 defined by

$$\pi_1 (e_1, e_2) = e_1, \quad \pi_2 (e_1, e_2) = e_2$$

and injections j_1, j_2 defined by

$$j_1(e_1) = (e_1, \sigma_2^\circ \circ p_1(e_1)), \quad j_2(e_2) = (\sigma_1^\circ \circ p_2(e_2), e_2);$$

here

$$\sigma_2^\circ \circ p_1(e_1) = 0 \quad \text{and} \quad \sigma_1^\circ \circ p_2(e_2) = 0.$$

Clearly, $\pi_1 \circ j_1 =$ identity and $\pi_2 \circ j_2 =$ identity. Moreover, the sequence

$$0 \longrightarrow E_1 \xrightarrow{j_1} E_1 \oplus E_2 \xrightarrow{\pi_2} E_2 \longrightarrow 0$$

is exact.

If now

$$0 \longrightarrow E_1 \xrightarrow{\phi} E_3 \xrightarrow{\eta} E_2 \longrightarrow 0$$

is any exact sequence of vector bundles and X-homomorphisms, there is an X-isomorphism $\alpha: E_3 \to E_1 \oplus E_2$ such that the following diagram

$$\begin{array}{ccccccccc}
0 & \longrightarrow & E_1 & \xrightarrow{\phi} & E_3 & \xrightarrow{\eta} & E_2 & \longrightarrow & 0 \\
 & & \| & & \downarrow{\scriptstyle\alpha} & & \| & & \\
0 & \longrightarrow & E_1 & \xrightarrow{j_1} & E_1 \oplus E_2 & \xrightarrow{\pi_2} & E_2 & \longrightarrow & 0
\end{array}$$

commutes; that is, $\alpha \circ \phi = j_1$ and $\eta \circ \alpha^{-1} = \pi_2$.

If E is an n-dimensional vector bundle over X, Y is any space, and f: $Y \to X$ is a continuous map, the induced fiber space f^*E is an n-dimensional vector bundle over Y.

9.3 THEOREM *Let $p_1 : E_1 \to X_1$ and $p_2 : E_2 \to X_2$ be two vector bundles. Let (ϕ, ψ): $E_1 \to E_2$ be a homomorphism. There are an X_1-homomorphism α: $E_1 \to \psi^*E_2$ and a bundle isomorphism (β, ψ): $\psi^*E_2 \to E_2$ such that $\beta \circ \alpha = \phi$.*

Proof The total space ψ^*E_2 is the subspace of $X_1 \times E_2$ consisting of those pairs (x_1, e_2) such that $\psi(x_1) = p_2(e_2)$. Define (β, ψ) by $(\beta, \psi)((x_1, e_2), x_1) = (\beta (x_1, e_2) = e_2, \ \psi(x_1))$. Define now α: $E_1 \to X_1 \times E_2$ by $\alpha(e_1)$

$= (p_1(e_1), \phi(e_1))$. The following diagram

$$\begin{array}{ccccc} E_1 & \xrightarrow{\alpha} & \psi^*E_2 & \xrightarrow{\beta} & E_2 \\ \downarrow{p_1} & & \downarrow & & \downarrow{p_2} \\ X_1 & = & X_1 & \xrightarrow{\psi} & X_2 \end{array}$$

shows that $(\phi, \psi) = (\beta, \psi) \circ \alpha$. ○

Let M_1 and M_2 be two differentiable manifolds. Let $\psi: M_1 \to M_2$ be a regular differentiable map. Then $(\psi_*, \psi): T(M_1) \to T(M_2)$ is a vector bundle homomorphism defined by $(\psi_*, \psi)(e, m) = (\psi_*(e), \psi(m))$ for $m \in M_1$ and $e \in T_m(M_1)$. By theorem 9.3 there are an M_1-homomorphism α: $T(M_1) \to \psi^*T(M_2)$ and a bundle isomorphism (β, ψ): $\psi^*T(M_2) \to T(M_2)$ such that $(\psi_*, \psi) = (\beta, \psi) \circ \alpha$.

Since ψ is a regular map, ψ_* is injective and therefore α is injective. Thus, Coker α is a well defined vector bundle called the *normal bundle* of ψ and denoted by $N(\psi)$. Then the sequence

$$0 \to T(M_1) \to \psi^*T(M_2) \to N(\psi) \to 0$$

is exact. It follows that the vector bundle $\psi^*T(M_2)$ induced by ψ on M_1 is isomorphic and, therefore, equivalent to the direct sum $T(M_1) \oplus N(\psi)$.

10.1 Problems

1 (1.1) Let $E = (E, X, p)$ be a fiber space, let $f: Y \to X$ be a continuous map and let $f^*E = (E', Y, \pi)$ be the induced fiber space. Prove that if p is an open map, π is also an open map.

2 (4.1) Show that the tangent bundle of a nonorientable manifold is never trivial.

3 (7.1) Let I be the interval $(0 < t < 1)$ and let $\phi: I \to M$ be a differentiable curve on a differentiable manifold M. If $ds^2 = \sum_{i,j} g_{ij}\, du^i du^j$ is the local expression of a Riemannian metric on M, find the expression of the induced metric on I.

4 (8.1) Let $E = \{(l, x) \in \mathbf{P}^{n-1}(\mathbf{R}) \times \mathbf{R}^n \mid x \in l\}$ and let $p: \mathbf{P}^{n-1}(\mathbf{R}) \times \mathbf{R}^n \to \mathbf{P}^{n-1}(\mathbf{R})$ be the natural projection. Prove that $(E, \mathbf{P}^{n-1}(\mathbf{R}), p)$ is a line bundle.

5 (8.1) If $E \to M$ is a vector bundle over M, denote by E_0 the set of all nonzero vectors of E. Let M be a connected n-dimensional differentiable manifold and let $A_0^n = \wedge^n T_0^*(M)$. Show that:

a) either A_0^n is connected or it has exactly two components,

b) M is orientable if and only if A_0^n has two components.

6 (8.1) Fill in all missing details in definitions 8.7, 8.8 and 8.10.

7 (8.1) Let E, F be two vector bundles over a topological space X and let $h : E \times F \to E \otimes F$ be the natural X-morphism. Show that the tensor product of vector bundles satisfies the following universal property: for every vector bundle G over X and bilinear X-homomorphism $f : E \times F \to G$ there is a unique X-homomorphism $\phi : E \otimes F \to G$ such that $\phi \circ h = f$.

8 (9.1) Let $X = [0, 1]$, let E be the trivial vector bundle $X \times \mathbf{R}$, and let $\phi : E \to E$ be defined by $\phi(x, t) = (x, xt)$. Show that Ker ϕ is not a vector bundle.

9 (9.1) Let $\phi : E_1 \to E_2$ be an X-homomorphism of vector bundles. Prove that if rank ϕ is constant, that is, if rank $\phi \mid E_{1x}$ is constant for every $x \in X$, then Ker ϕ and Coker ϕ are vector bundles.

CHAPTER 7

Sheaves

1.1 Presheaves

LET X BE a topological space and let $\{U\}$ be the collection of all open sets of X.

1.2 DEFINITION *A presheaf of sets over X consists of:*

i) *a collection of sets* $\{A_U\}_{U \in \{U\}}$,

ii) *for every pair* (U, V) *of open sets with* $V \subset U$, *a map* $\varrho_V^U \colon A_U \to A_V$ *such that:*

(1) *for every triple* U, V, W *of open sets with* $W \subset V \subset U$,

$$\varrho_W^V \circ \varrho_V^U = \varrho_W^U,$$

(2) *for every* $U \in \{U\}$,

$$\varrho_U^U = \textit{identity}.$$

We shall denote a presheaf by $A = \{A_U, \varrho_V^U\}$; the maps of the collection $\{\varrho_V^U\}$ will be called the *restriction maps* of the presheaf.

Examples of presheaves of sets are given by the cross-sections of a fiber space, which we considered in section 6.1.1. If $p \colon E \to X$ is a fiber space, then $\{\Gamma(U, E), \varrho_V^U\}$ is a presheaf of sets over X.

If $\{A_U, \varrho_V^U\}$ is a presheaf over X such that, for every U, the set A_U is an abelian group and, for every pair (U, V) with $V \subset U$, the map $\varrho_V^U \colon A_U \to A_V$ is a group homomorphism, then $\{A_U, \varrho_V^U\}$ is called a *presheaf of abelian groups* over X.

In a similar way, we define presheaves of rings, modules, vector spaces and so on.

We will now give some other examples of presheaves over the topological space X.

1.3 Let Y be a set; let $A_U = \mathrm{Hom}(U, Y)$ be the set of all maps from the open set $U \subset X$ into Y, and for every $V \subset U$ let $\{\varrho_V^U\}$ be the natural restrictions to V of these maps. Then $\{A_U, \varrho_V^U\}$ is a presheaf of sets over X.

1.4 If Y is a topological space, we can consider the set $A_U = \mathrm{Hom}_c(U, Y)$ of all continuous maps $U \to Y$. With the same definition of $\{\varrho_V^U\}$ as before, $\{A_U, \varrho_V^U\}$ is a presheaf of sets over X.

1.5 If in the previous example we assume that Y has discrete topology, the sets A_U are the sets of locally constant maps $X \to Y$. Then $\{A_U, \varrho_V^U\}$ is said to be a *locally constant* presheaf.

1.6 If X and Y are complex manifolds, we can consider the set $A_U = \text{Hom}_0\,(U, Y)$ of all holomorphic maps $U \to Y$. Then $\{A_U, \varrho_V^U\}$ is the presheaf over X of holomorphic maps from open subsets of X into Y.

1.7 Let Y be a topological group. Then $A_U = \text{Hom}_c\,(U, Y)$ has a natural group structure if we define

$$fg : U \to Y$$

by

$$fg\,(x) = f(x)\,g(x)$$

for $x \in U$; $f, g \in \text{Hom}_c\,(U, Y)$. Then $\{A_U, \varrho_V^U\}$ is a presheaf of abelian groups over X.

1.8 In particular, if $Y = \mathbf{R}$ (or $\mathbf{C}$), by considering $A_U = \text{Hom}_c\,(U, \mathbf{R})$ (or $= \text{Hom}_c\,(U, \mathbf{C})$) as a group under addition, the above example gives the presheaf over X of real valued (or complex valued) continuous functions. It is a presheaf of rings over X.

1.9 If X is a complex manifold and $Y = \mathbf{C}$, we can define the presheaf over X of holomorphic functions on open subsets of X.

2.1 Presheaf homomorphisms

Let $A = \{A_U, \varrho_V^U\}$ and $A' = \{A'_U, \varrho'^U_V\}$ be two preheaves of sets over the same topological space X. A *presheaf map*

$$\phi : A \to A'$$

is a collection $\{\phi_U\}$ of maps

$$\phi_U : A_U \to A'_U$$

such that, for every pair (U, V) with $V \subset U$, the following diagram

$$\begin{array}{ccc} A_U & \xrightarrow{\phi_U} & A'_U \\ {\scriptstyle \varrho_V^U}\downarrow & & \downarrow{\scriptstyle \varrho'^U_V} \\ A_V & \xrightarrow{\phi_V} & A'_V \end{array}$$

commutes.

A presheaf map $\phi: A \to A'$ is said to be *injective* (respectively *surjective*) if for every U the map $\phi_U: A_U \to A'_U$ is injective (respectively surjective).

We will give some examples of presheaf maps which frequently occur.

Let X, Y be two topological spaces and let $A = \{A_U, \varrho_V^U\}$ be the presheaf of continuous maps of open subsets of X into Y. Let $A' = \{A'_U, \varrho'^U_V\}$ be the presheaf of arbitrary maps of open subsets of X into Y. For every U we have an obvious map

$$\phi_U : A_U \to A'_U .$$

The collection $\{\phi_U\}$ of these maps, which clearly commute with the restriction maps of the two presheaves, defines an injective presheaf map

$$\phi : A \to A'.$$

If X is a complex manifold, there is a presheaf map of the presheaf of holomorphic functions over X into the presheaf of continuous functions over X.

If the presheaves A and A' over X are presheaves of abelian groups, a presheaf map $\phi: A \to A'$ is called a *presheaf homomorphism* if for every U the map $\phi_U: A_U \to A'_U$ is a group homomorphism. A presheaf homomorphism $\phi = \{\phi_U\}$ is called a presheaf isomorphism if, for every U, ϕ_U is an isomorphism. Then $\phi^{-1} = \{\phi_U^{-1}\}$ is a presheaf homomorphism, $\phi \circ \phi^{-1} = \text{id.}$, $\phi^{-1} \circ \phi = \text{id.}$

There are similar definitions when A, A' are presheaves of rings, vector spaces and so on.

Let A and A' be presheaves of abelian groups over X. If for every U the abelian group A_U is a subgroup of A'_U and for every $V \subset U$ the group homomorphism ϱ_V^U is the restriction to A_U of the group homomorphism $\varrho'^U_V: A'_U \to A'_V$, then A is said to be a *subpresheaf* of A. If we consider now the canonical inclusions

$$i_U : A_U \to A'_U,$$

we clearly obtain a presheaf homomorphism $i = \{i_U\}: A \to A'$ called the *embedding* of A into A'.

If for every U the abelian group A'_U is a quotient of the group A_U and for every $V \subset U$ the group homomorphism ϱ'^U_V is the homomorphism induced from ϱ_V^U, the presheaf A' is called a *quotient presheaf* of A. If we now consider the projection homomorphisms

$$p_U : A_U \to A'_U,$$

we clearly obtain a presheaf homomorphism $p = \{p_U\}: A \to A'$.

Let A, A' be presheaves over X and let $\phi\colon A \to A'$ be a presheaf homomorphism. The kernel of ϕ is a presheaf over X, defined by $\operatorname{Ker}\phi = \{\operatorname{Ker}\phi_U, \varrho^U_V\}$. It is clear that $\operatorname{Ker}\phi$ is a subpresheaf of A. Similarly, the image of ϕ is a presheaf over X, defined by $\operatorname{Im}\phi = \{\operatorname{Im}\phi_U, \varrho'^U_V\}$. $\operatorname{Im}\phi$ is a subpresheaf of A'.

Let $\{A^n\} = \{A^n_U, \varrho^{nU}_V\}$ be a sequence of presheaves of abelian groups over the same topological space X and let $\{\phi^n\}$ be a sequence of presheaf homomorphisms. For every n, $\phi^n\colon A^n \to A^{n+1}$ is a collection $\{\phi^n_U\}$ of homomorphisms $\phi^n_U\colon A^n_U \to A^{n+1}_U$.

2.2 DEFINITION *The sequence*

$$\cdots \to A^{n-1} \xrightarrow{\phi^{n-1}} A^n \xrightarrow{\phi^n} A^{n+1} \to \cdots$$

of presheaves and presheaf homomorphisms is said to be an exact sequence of presheaves if for every U the sequence of abelian groups and group homomorphisms

$$\cdots \to A^{n-1}_U \xrightarrow{\phi^{n-1}_U} A^n_U \xrightarrow{\phi^n_U} A^{n+1}_U \to \cdots$$

is exact.

We give now some examples of exact sequences of presheaves.

2.3 Let A be a presheaf, let B be a subpresheaf of A and let A/B be the quotient presheaf. Then the sequence

$$0 \to B \xrightarrow{i} A \xrightarrow{p} A/B \to 0$$

is an exact sequence of presheaves and presheaf homomorphisms. Here 0 denotes a presheaf which, to every U, associates a group consisting only of the neutral element.

2.4 Let X be a topological space and let G be any group considered as a topological space with discrete topology. Let us denote by the same letter G the presheaf of locally constant maps from X into the group G as in example 1.5.

Given any exact sequence of groups and group homomorphisms

$$\cdots \to G^{n-1} \xrightarrow{\phi^{n-1}} G^n \xrightarrow{\phi^n} G^{n+1} \to \cdots,$$

the corresponding sequence of locally constant presheaves over X and induced presheaf homomorphisms

$$\cdots \to G^{n-1} \xrightarrow{\phi^{n-1}} G^n \xrightarrow{\phi^n} G^{n+1} \to \cdots$$

is also exact.

2.5 As a special case, let us consider the exact sequence of groups and group homomorphisms

$$0 \to \mathbf{Z} \xrightarrow{n} \mathbf{Z} \to \mathbf{Z}_n \to 0$$

where **Z** is the group of the integers, the homomorphism n is defined by $n(z) = nz$ and $\mathbf{Z}_n = \mathbf{Z}/n\mathbf{Z}$. Then, for every topological space X, we have an exact sequence of presheaves over X,

$$0 \to \mathbf{Z} \xrightarrow{n} \mathbf{Z} \to \mathbf{Z}_n \to 0,$$

with the obvious induced presheaf homomorphisms.

2.6 As a second special case, let $\mathbf{C}^*$ be the multiplicative group of **C** and let

$$e^{2\pi i} : \mathbf{C} \to \mathbf{C}^*$$

be the homomorphism defined by $e^{2\pi i}(z) = e^{2\pi i z}$. Since $e^{2\pi i z} = 1$ if and only if $z \in \mathbf{Z}$, then $\operatorname{Ker} e^{2\pi i} = \mathbf{Z}$. We then have an exact sequence of groups and group homomorphisms

$$0 \to \mathbf{Z} \xrightarrow{j} \mathbf{C} \xrightarrow{e^{2\pi i}} \mathbf{C}^* \to 0$$

where $j: \mathbf{Z} \to \mathbf{C}$ is the natural embedding. Then, for every topological space X, we have an induced sequence of locally constant presheaves over X

$$0 \to \mathbf{Z} \xrightarrow{j} \mathbf{C} \xrightarrow{e^{2\pi i}} \mathbf{C}^* \to 0$$

which is also exact.

2.7 We now give a third example. Let **Z** be the locally constant presheaf of integers over X as before, let $\mathbf{C}_c$ be the presheaf over X of continuous maps of open subsets of X into **C** and let $\mathbf{C}_c^*$ be the presheaf over X of continuous maps of open subsets of X into $\mathbf{C}^*$. Let

$$\phi : \mathbf{C}_c \to \mathbf{C}_c^*$$

be the presheaf homomorphism defined for every U by

$$\phi_U : \alpha \to e^{2\pi i \alpha}$$

where

$$\alpha : U \to \mathbf{C}$$

is a continuous map. If α is such that $e^{2\pi i \alpha} = 1$, then the value of α must be an integer. It follows that $\operatorname{Ker} \phi_U = \mathbf{Z}$. Thus we have the exact sequence of presheaves

$$0 \to \mathbf{Z} \to \mathbf{C}_c \to \mathbf{C}_c^* .$$

We shall show that, in general, the sequence

$$0 \to \mathbf{Z} \to \mathbf{C}_c \to \mathbf{C}_c^* \to 0$$

is not exact. Let $X = \mathbf{R}^2$ and let $U = \{(x, y) \in \mathbf{R}^2 \mid 0 < x^2 + y^2 < 1\}$. Let $\alpha^*: U \to \mathbf{C}^*$ be the function $(x, y) \to z = x + iy$. If there were a function $\alpha: U \to \mathbf{C}$ such that $e^{2\pi i\alpha} = z$, then we would have $\alpha = (1/2\pi i) \log z$, which is not a single valued function.

3.1 Sheaves

In section 6.2.1 we gave the definition of a sheaf of sets. If $p: \mathscr{E} \to X$ is a sheaf of sets, the projection p satisfying conditions (i), (ii) of definition 6.2.2 will be called a *local homeomorphism*. The map p is an open map. If $\mathscr{E}$ is a sheaf of sets over X, the set $\mathscr{E}_x = p^{-1}(x)$ is called the *stalk* of the sheaf $\mathscr{E}$ over $x \in X$.

A sheaf $p: \mathscr{E} \to X$ is called a *constant sheaf* if there exists a topological space Y with discrete topology such that $\mathscr{E} = X \times Y$ and $\mathscr{E}$ has the product topology. Then, if $\mathscr{E}$ is a constant sheaf over X, we have $\mathscr{E}_x = Y$ for every $x \in X$; moreover, for every open set $U \subset X$ the cross-sections of $\mathscr{E}$ over U are constant.

If $p: \mathscr{E} \to X$ is any sheaf, we shall denote by $\mathscr{E} \times_X \mathscr{E}$ the subspace of the topological product $\mathscr{E} \times \mathscr{E}$ consisting of all pairs of points of $\mathscr{E}$ belonging to the same stalk, that is,

$$\mathscr{E} \times_X \mathscr{E} = \{(e, e') \in \mathscr{E} \times \mathscr{E} \mid p(e) = p(e')\}.$$

3.2 DEFINITION *A sheaf of sets $\mathscr{E}$ is called a sheaf of abelian groups if*

i) *for every $x \in X$ the stalk $\mathscr{E}_x$ is an abelian group,*

ii) *the map*

$$\mathscr{E} \times_X \mathscr{E} \to \mathscr{E}$$

is continuous.

We remark that the map $\mathscr{E} \times_X \mathscr{E} \to \mathscr{E}$ which sends (e, e') into $e - e'$ is a well defined operation since both e and e' belong to the same stalk and every stalk has the structure of an abelian group. Condition (ii) requires that this operation be continuous.

Another way of looking at this continuity condition is the following: Let U be an open subset of X and let $\Gamma(U, \mathscr{E})$ be the set of cross-sections of $\mathscr{E}$ over U. If $\sigma, \sigma' \in \Gamma(U, \mathscr{E})$, condition (ii) implies that $\sigma - \sigma' \in \Gamma(U, \mathscr{E})$. Moreover, the zero cross-section $\sigma^0: U \to \mathscr{E}$, which to every $x \in U$ associates the zero element of the group $\mathscr{E}_x$, clearly belongs to $\Gamma(U, \mathscr{E})$. It follows that $\Gamma(U, \mathscr{E})$ has a natural group structure. It is clear that the existence of this group structure on $\Gamma(U, \mathscr{E})$ for every $U \subset X$ is equivalent to condition (ii).

Sheaves of rings, vector spaces, and so on are defined in a similar way.

If $p: \mathscr{E} \to X$ and $q: \mathscr{F} \to X$ are two sheaves over X, an X-morphism $\phi: \mathscr{E} \to \mathscr{F}$ is called a *sheaf map*. If $\mathscr{E}, \mathscr{F}$ are sheaves of abelian groups over X and if the restriction $\phi_x = \phi \mid \mathscr{E}_x: \mathscr{E}_x \to \mathscr{F}_x$ is a group homomorphism for every $x \in X$, then ϕ is called a *sheaf homomorphism*. A sheaf homomorphism is called a *sheaf isomorphism* if it is an isomorphism between corresponding stalks.

A sheaf $\mathscr{E}$ over X is said to be a *subsheaf* of a sheaf $\mathscr{F}$ over X if $\mathscr{E}$ is an open subset of $\mathscr{F}$ and for every $x \in X$ the group $\mathscr{E}_x$ is a subgroup of $\mathscr{F}_x$.

If $\mathscr{E}$ is a sheaf over X, it follows from theorem 6.2.3 that the cross-sections of $\mathscr{E}$ over the open sets of X are a basis for the open sets of $\mathscr{E}$. It follows that, if $\phi: \mathscr{E} \to \mathscr{F}$ is a sheaf map, then ϕ is an open map. In fact, let U be any open set of X and let $\sigma \in \Gamma(U, \mathscr{E})$; then $\phi \circ \sigma \in \Gamma(U, \mathscr{F})$. Now since the sets $\sigma(U)$ and $\phi \circ \sigma(U)$ are open in $\mathscr{E}$ and $\mathscr{F}$ respectively, ϕ is an open map. It follows that the image of ϕ, $\phi(\mathscr{E})$ is a subsheaf of $\mathscr{F}$.

We also have that the inverse image under ϕ of the zero cross-section of $\mathscr{F}$ is an open set in $\mathscr{E}$, since every cross-section is an open set. It follows that $\mathscr{K}er\ \phi$ is a subsheaf of $\mathscr{E}$.

If $\mathscr{E}$ is a subsheaf of $\mathscr{F}$, we can define the *quotient sheaf* $\mathscr{F}/\mathscr{E}$. We leave all details to the reader, together with the proof that the sequence

$$0 \to \mathscr{E} \to \mathscr{F} \to \mathscr{F}/\mathscr{E} \to 0$$

is exact. Here 0 denotes a sheaf whose stalk over every point consists only of the neutral element.

If $p: \mathscr{E} \to X$ and $q: \mathscr{F} \to X$ are two sheaves of abelian groups over the same space X, we can consider a new sheaf $\pi: \mathscr{E} \times \mathscr{F} \to X \times X$ by defining the projection π by $\pi(e, f) = (p(e), q(f))$. This is clearly a sheaf of abelian groups. Therefore,

$$\mathscr{E} \underset{X}{\times} \mathscr{F} = \{(e, f) \in \mathscr{E} \times \mathscr{F} \mid p(e) = q(f)\}$$

is a well defined sheaf of abelian groups over X as the diagonal of $X \times X$. For every $x \in X$, the stalk over x is the abelian group $\mathscr{E}_x \times \mathscr{F}_x$.

3.3 Definition *Let $\mathscr{A}$ be a sheaf of rings over X and let $\mathscr{E}$ be a sheaf of abelian groups over X. If there is a sheaf homomorphism*

$$\phi : \mathscr{A} \underset{X}{\times} \mathscr{E} \to \mathscr{E}$$

such that, for every $x \in X$, the homomorphism ϕ_x: $\mathscr{A}_x \times \mathscr{E}_x \to \mathscr{E}_x$ gives $\mathscr{E}_x$ the structure of an $\mathscr{A}_x$-module, $\mathscr{E}$ is called a sheaf of modules over a sheaf of rings or simply an $\mathscr{A}$-Module.

A special case of an *$\mathscr{A}$-Module* occurs when the sheaf of rings $\mathscr{A}$ is a constant sheaf. We shall consider only the case when $\mathscr{A}$ is a sheaf of commutative rings with unity.

4.1 Sheaves of germs

We shall now develop a technique which relates the theory of fiber spaces and, in particular, sheaves, to that of presheaves.

We already know that, if E is a fiber space over X, the cross-sections of E over the open sets of X form a presheaf $\{\Gamma(U, E), \varrho_V^U\}$.

Assume now that we have a presheaf $A = \{A_U, \varrho_V^U\}$ over X. For every $x \in X$, the family $\mathscr{U}_x$ of all open sets of X containing the point x is partially ordered by inclusion. Moreover, if $W, V, U \in \mathscr{U}_x$ and $W \subset V$, $V \subset U$, then $W \subset U$. To every $U \in \mathscr{U}_x$ there corresponds a set A_U and for every pair (U, V) with $V \subset U$ there is a restriction map ϱ_V^U: $A_U \to A_V$ such that, for every $U \in \mathscr{U}_x$, ϱ_U^U = identity and for every $W \subset V \subset U$,

$$\varrho_W^V \circ \varrho_V^U = \varrho_W^U.$$

We shall denote by a_U an element of A_U and let

$$\tilde{A}_x = \bigcup_{\substack{a_U \in A_U \\ U \in \mathscr{U}_x}} a_U.$$

We identify two elements $a_U, a_V \in \tilde{A}_x$ if there exists a $W \in \mathscr{U}_x$ with $W \subset U$, $W \subset V$, such that $\varrho_W^U a_U = \varrho_W^V a_V$. This is an equivalence relation, as it follows at once from the partial ordering and from the properties of the restriction maps ϱ_V^U.

The set of equivalence classes so obtained is called the *direct limit* of the system $\{A_U, \varrho_V^U\}$ and is denoted by

$$A_x = \varinjlim \{A_U, \varrho_V^U\}.$$

For every $U \in \mathscr{U}_x$ we shall denote by ϱ_x^U the natural map $A_U \to A_x$ which sends any element a_U into its equivalence class in A_x.

If for every $U \in \mathscr{U}_x$ the set A_U is an abelian group and every ϱ_V^U is a group homomorphism, it is easy to see that A_x has a natural abelian group structure.

We shall now show that the collection of the A_x defines a sheaf $\mathscr{A}$ over X whose stalks are the A_x. We begin by defining $\mathscr{A} = \bigcup_{x \in X} A_x$. An element $\alpha \in \mathscr{A}$ is called a *germ* of the presheaf A. A representative of a germ consists of:

i) a point $x \in X$ called the *center* of the germ,

ii) an open set U of X containing x,

iii) an element a_U of the set A_U associated to the open set U.

All representatives of the same germ have the same center.

We now define a projection $p: \mathscr{A} \to X$ by $p(A_x) = x$. The map p is clearly surjective.

The last step is to give $\mathscr{A}$ a suitable topology so that p is a local homeomorphism. Let U be any open set of X and let $a \in A_U$. This element is a representative of a germ α for every $x \in U$. Denote by $\Omega(U, a)$ the set of all these germs. It is easy to see that, letting U run over all open sets of X and a over all elements of the corresponding A_U, we obtain a basis for a topology in $\mathscr{A}$. In fact, if $a \in A_U$ and $a' \in A_{U'}$ are such that $\varrho_x^U a = \varrho_x^{U'} a'$, it follows that x belongs to some open set V of X with $V \subset U \cap U'$ and $\varrho_V^U a = \varrho_V^{U'} a'$. Hence, $\Omega(V, \varrho_V^U a) \subset \Omega(U, a) \cap \Omega(U', a')$.

When $\mathscr{A}$ is endowed with this topology, every open set $\Omega(U, a)$ is homeomorphic to the open set U for every open set U of X under the projection p. Thus $p: \mathscr{A} \to X$ is a sheaf of sets over X. The sheaf $\mathscr{A}$ is called the *sheaf of germs of the presheaf A*.

If A is a presheaf of abelian groups, it is an easy matter to show that the sheaf of germs $\mathscr{A}$ associated to A is a sheaf of abelian groups, that is, to show that the group operations are continuous in the sense of definition 3.2. It is also clear that the sheaf of germs of a presheaf of rings is a sheaf of rings.

Let $A = \{A_U, \varrho_V^U\}$ be a presheaf of rings and $E = \{E_U, \sigma_V^U\}$ be a presheaf of abelian groups over X such that every E_U is an A_U-module and the restriction maps are compatible with the module structures. From the above construction applied to the presheaves A and E we obtain an $\mathscr{A}$*-Module* $\mathscr{E}$; that is, the sheaf $\mathscr{E}$ of germs of the presheaf E is a sheaf of modules over the sheaf of rings $\mathscr{A}$, the sheaf of germs of the presheaf A.

If $\mathscr{A}$ and $\mathscr{B}$ are two sheaves of sets over X we shall denote by Hom $(\mathscr{A}, \mathscr{B})$ the set of X-morphisms $\alpha: \mathscr{A} \to \mathscr{B}$. If U is an open subset of X, for every $\alpha \in \text{Hom}(\mathscr{A}, \mathscr{B})$ there is an induced map $\mathscr{A} \mid U \to \mathscr{B} \mid U$. If now $V \subset U$, there is an obvious map

$$\varrho_V^U : \text{Hom}(\mathscr{A} \mid U, \mathscr{B} \mid U) \to \text{Hom}(\mathscr{A} \mid V, \mathscr{B} \mid V)$$

and it is easy to see that $\{\text{Hom}(\mathscr{A} \mid U, \mathscr{B} \mid U), \varrho_V^U\}$ is a presheaf of sets over X. The sheaf of germs of this presheaf is called *the sheaf of germs of X-morphisms from* $\mathscr{A}$ *to* $\mathscr{B}$ and is denoted by $\mathscr{H}om(\mathscr{A}, \mathscr{B})$.

If $\mathscr{E}, \mathscr{F}$ are two $\mathscr{A}$*-Modules* over X, we can define the sheaf of germs of homomorphisms from $\mathscr{E}$ to $\mathscr{F}$ denoted by $\mathscr{H}om_{\mathscr{A}}(\mathscr{E}, \mathscr{F})$, and it is easy to show that $\mathscr{H}om_{\mathscr{A}}(\mathscr{E}, \mathscr{F})$ is a sheaf of abelian groups. This sheaf is not, in general, an $\mathscr{A}$*-Module*; however, if $\mathscr{A}$ is a sheaf of commutative rings, then it is an $\mathscr{A}$*-Module* in a natural way.

We shall now give some examples of sheaves of germs.

4.2 *Sheaf of germs of continuous functions* Let X be a topological space. We consider the fiber space $p: X \times \mathbf{R} \to X$ and the presheaf $\{\Gamma(U, X \times \mathbf{R}), \varrho_V^U\}$ of cross-sections associated to that fiber space. The sheaf of germs of this presheaf is called the *sheaf of germs of real valued continuous functions on X.*

4.3 *Sheaf of germs of differentiable cross-sections with values in a differentiable fiber bundle* $p: E \to M$ This sheaf is the sheaf $\mathscr{E}$ of germs of the presheaf $\{\Gamma_d(U, E), \varrho_V^U\}$ of differentiable cross-sections of E over open sets of the differentiable manifold M.

Particular cases of this sheaf occur when E is a differentiable vector bundle over M. More particularly, if E is a trivial line bundle, the sheaf $\mathscr{E}$ is the *sheaf of germs of real valued differentiable functions on M.*

Sheaves of germs of differential forms of any degree are defined this way. For every r, $0 \leq r \leq \dim_{\mathbf{R}} M$, $A^r(M)$ is a real differentiable vector bundle orer M. The sheaf $\mathscr{A}^r(M)$ of germs of the presheaf $\{\Gamma_d(U, A^r(M)), \varrho_V^U\}$ of differentiable cross-sections of $A^r(M)$ is called the *sheaf of germs of differential forms of degree r on M.*

4.4 *Sheaf of germs of holomorphic sections with values in a holomorphic fiber bundle $p: E \to M$* This sheaf is the sheaf $\mathscr{E}$ of germs of the presheaf $\{\Gamma_0(U, E), \varrho_V^U\}$ of holomorphic cross-sections of E over open sets of the complex manifold M.

In particular, if E is a trivial line bundle, the sheaf $\mathscr{E}$ of its holomorphic cross-sections is the sheaf of germs of holomorphic functions on M.

Also, for every r, s, $0 \leq r, s \leq \dim_{\mathbf{C}} M$, we can define the *sheaf* $\mathscr{E}^{r,s}(M)$ *of germs of differential forms of type* (r, s) *on* M.

4.5 *Remark* Let $\mathscr{E}$ be a sheaf over X. In general, the topology of $\mathscr{E}$ is not Hausdorff even if we assume that X is a Hausdorff space. In fact, let X be a Hausdorff space, let U be an open set of X and let σ, τ be two elements of $\Gamma(U, \mathscr{E})$. The set of all $x \in U$ such that $\sigma(x) = \tau(x)$ is open in U. If, now, $\mathscr{E}$ is Hausdorff, this set is also closed in U; that is, if two cross-sections σ, τ have the same value at one point $x \in U$, they coincide on the connected component of U containing x. In other words, a Hausdorff sheaf satisfies the "principle of analytic continuation".

It is easy to check that the sheaf of germs of holomorphic functions on a complex manifold M is Hausdorff.

On the other hand, the sheaf of germs of continuous functions on a Hausdorff space X, in general, is not Hausdorff.

For example, let $X = \mathbf{R}$, and consider the functions

$$f(x) = \begin{cases} 0 & \text{if } x < 0 \\ x & \text{if } x \geq 0 \end{cases}$$

and $g(x) = 0$.

The two germs $\mathscr{f}_0, \mathscr{g}_0$ at $x = 0$, defined by $\mathscr{f}_0 = \varrho_0^X f$, $\mathscr{g}_0 = \varrho_0^X g$, do not have disjoint neighborhoods.

5.1 Canonical presheaves

If $A = \{A_U, \varrho_V^U\}$ is a presheaf of sets over a space X, we associate to A its sheaf of germs $\mathscr{A}$. We can now consider the presheaf $\Gamma(\mathscr{A}) = \{\Gamma(U, \mathscr{A}), \sigma_V^U\}$ of cross-sections of $\mathscr{A}$. There is a natural presheaf map

$$\alpha : A \to \Gamma(\mathscr{A}),$$

defined by the natural maps

$$\alpha_U : A_U \to \Gamma(U, \mathscr{A}).$$

In fact, every element $a \in A_U$ defines an open set $\Omega(U, a)$ of $\mathscr{A}$ and, therefore, by theorem 6.2.3, a cross-section of $\mathscr{A}$ over U. We shall now investigate the properties of the map α.

5.2 DEFINITION *A presheaf $A = \{A_U, \varrho_V^U\}$ over a topological space X is called a canonical presheaf if for any family $\{U_i\}_{i \in I}$ of open sets of X with $U = \bigcup_{i \in I} U_i$ the two following conditions are satisfied:*

i) *if $a', a'' \in A_U$ such that for every $i \in I$*

$$\varrho_{U_i}^U a' = \varrho_{U_i}^U a''$$

then $a' = a''$,

ii) *if $a_i \in A_{U_i}$ are such that for every $i, j \in I$*

$$\varrho_{U_i \cap U_j}^{U_i} a_i = \varrho_{U_i \cap U_j}^{U_j} a_j$$

there exists an element $a \in A_U$ such that for every $i \in I$

$$\varrho_{U_i}^U a = a_i.$$

We remark that, if A is a canonical presheaf, the element $a \in A_U$, whose existence is stated by condition (ii), is unique by condition (i). The importance of canonical presheaves is shown by the next theorem.

5.3 THEOREM *Let A be a canonical presheaf over X and let $\mathscr{A}$ be the sheaf of germs of A. The presheaf map*

$$\alpha : A \to \Gamma(\mathscr{A})$$

is bijective.

Proof That α_U is injective follows at once from condition (i) of definition 5.2. In fact, if $a, a' \in A_U$ define the same cross-section $\sigma \in \Gamma(U, \mathscr{A})$, there exists an open set V such that $\varrho_x^V a = \varrho_x^V a'$ for every $x \in V$. It follows that U can be considered as the union of open sets U_i such that

$$\varrho_{U_i}^U a = \varrho_{U_i}^U a'$$

for every i and this implies that $a = a'$.

Let now $\sigma \in \Gamma(U, \mathscr{A})$. For every $x \in U$, $\sigma(x) \in \mathscr{A}_x$. This means that there exists an open set $\Omega(U_i, \sigma_i)$ of $\mathscr{A}$ such that $\sigma_i \in A_{U_i}$ and $\varrho_x^{U_i} \sigma_i = \sigma(x)$. We can consider U as the union of these open sets U_i. Since, on every intersection $U_i \cap U_j$ we have $\sigma_i = \sigma_j$, the existence of an element $\sigma_U \in A_U$ such that $\alpha_U(\sigma_U) = \sigma$ follows from condition (ii). Thus α is surjective. ○

If the canonical presheaf A is a presheaf of abelian groups, then, as it is easy to check, α is a presheaf isomorphism.

The above theorem shows that there is a one to one correspondence between sheaves and canonical presheaves. Most of the presheaves which practically occur are canonical presheaves. However, if $\mathscr{A}$ is any sheaf, the presheaf of its cross-sections is a canonical presheaf. In other words, we have:

5.6 THEOREM *Every sheaf is the sheaf of germs of some presheaf.* ○

6.1 Alexander–Spanier cochains

Let X be a topological space, let U be an open set of X and let G be an abelian group. For every integer $q \geq 0$, put

$$C_U^q = \text{Hom}\,(U^{q+1}, G)$$

where $\text{Hom}\,(U^{q+1}, G)$ denotes the set of maps $f(x_0, \dots, x_q)\colon U^{q+1} \to G$. For every U the set C_U^q has a natural abelian group structure. If $V \subset U$, a homomorphism $\varrho_V^U\colon C_U^q \to C_V^q$ is defined by restriction. We then have a presheaf of abelian groups over X,

$$C^q = \{C_U^q, \varrho_V^U\},$$

called the *Alexander–Spanier presheaf of q-dimensional cochains of X with values in G.*

We shall show that, for $q > 0$, the presheaf C^q is not canonical. Let $D = \{(x_0, \dots, x_q) \in U^{q+1} \mid x_0 = x_1 = \cdots = x_q\}$ be the diagonal of U^{q+1}. Let $f \in \text{Hom}\,(U^{q+1}, G)$ such that $f = 0$ on an open set $U(D)$ containing D. Let $\{U_i\}_{i \in I}$ be an open covering of U such that for every $i \in I$,

$$U_i^{q+1} \subset U(D).$$

Then we have $f = 0$ on U_i for every $i \in I$ but not necessarily $f = 0$ on U.

It is easy to check that C^0 is a canonical presheaf.

We now define a *coboundary* presheaf homomorphism

$$\delta^q : C^q \to C^{q+1}$$

as follows: if $f \in \text{Hom}\,(U^{q+1}, G)$ we put

$$(\delta_U^q f)\,(x_0, \dots, x_{q+1}) = \sum_{i=0}^{q+1} (-1)^i f(x_0, \dots, \widehat{x_i}, \dots, x_{q+1});$$

then $\delta_U^q f \in \operatorname{Hom}(U^{q+2}, G)$ and it is easy to check that $\delta^q = \{\delta_U^q\}$ is a presheaf map; that is, it commutes with the restriction maps. It is also trivial to verify that

$$\delta^{q+1} \circ \delta^q = 0.$$

We consider now the sequence of presheaves and presheaf homomorphisms

$$C^0 \xrightarrow{\delta^0} C^1 \xrightarrow{\delta^1} C^2 \xrightarrow{\delta^2} \cdots C^q \xrightarrow{\delta^q} \cdots$$

and look at $\operatorname{Ker} \delta^0$. If $f\colon U \to G$ is an element of $\operatorname{Hom}(U, G)$ then by definition

$$(\delta_U^0 f)(x_0, x_1) = f(x_1) - f(x_0).$$

Now if $f(x_1) - f(x_0) = 0 \in \operatorname{Hom}(U^2, G)$, this means that f is constant on every connected component of U. Thus, if we denote by G the presheaf of locally constant maps from X to G we obtain a sequence

6.2 $$0 \longrightarrow G \xrightarrow{\varepsilon} C^0 \xrightarrow{\delta^0} C^1 \xrightarrow{\delta^1} C^2 \xrightarrow{\delta^2} \cdots$$

of presheaves and presheaf homomorphisms where the presheaf map ε is injective. This map ε is called an *augmentation*. Clearly $\delta^0 \circ \varepsilon = 0$. The sequence 6.2 is called an *augmented cochain complex of presheaves*. We have seen that it satisfies the following properties:

$$\delta^0 \circ \varepsilon = 0, \quad \delta^{q+1} \circ \delta^q = 0;$$

moreover, the sequence

$$0 \to G \xrightarrow{\epsilon} C^0 \xrightarrow{\delta^0} C^1$$

is exact.

7.1 Induced sequences

Every presheaf homomorphism

$$\phi : E \to F$$

induces in a natural way a sheaf homomorphism

$$\varphi : \mathscr{E} \to \mathscr{F}$$

between sheaves of germs. If

$$0 \to E \xrightarrow{\alpha} F \xrightarrow{\beta} G \to 0$$

is an exact sequence of presheaves and presheaf homomorphisms, it is easy to check that the associated sequence of sheaves of germs and sheaf homomorphisms

$$0 \to \mathscr{E} \xrightarrow{\alpha} \mathscr{F} \xrightarrow{\beta} \mathscr{G} \to 0$$

is also exact.

Conversely, if $\mathscr{E}$, $\mathscr{F}$ are two sheaves over X, every sheaf homomorphism

$$\varphi : \mathscr{E} \to \mathscr{F}$$

induces a presheaf homomorphism

$$\phi : \Gamma(\mathscr{E}) \to \Gamma(\mathscr{F})$$

between the presheaves of their cross-sections by the natural group homomorphisms

$$\phi_U : \Gamma(U, \mathscr{E}) \to \Gamma(U, \mathscr{F}).$$

It is now easy to verify that, if φ is injective, ϕ is also injective, while, if φ is surjective, ϕ does not need to be so. It follows that if

$$0 \to \mathscr{E} \xrightarrow{\alpha} \mathscr{F} \xrightarrow{\beta} \mathscr{G} \to 0$$

is an exact sequence of sheaves and sheaf homomorphisms, the associated sequence of cross-sections

$$0 \to \Gamma(\mathscr{E}) \xrightarrow{\alpha} \Gamma(\mathscr{F}) \xrightarrow{\beta} \Gamma(\mathscr{G}) \to 0$$

in general is not exact because β is not necessarily surjective. However, we remark that the sequence

$$0 \to \Gamma(\mathscr{E}) \xrightarrow{\alpha} \Gamma(\mathscr{F}) \xrightarrow{\beta} \Gamma(\mathscr{G})$$

is always exact.

As an example, we consider the following sheaves over a manifold M: the constant sheaf 0, the sheaf of integers $\mathscr{Z}$, the sheaf of germs of complex valued continuous functions $\mathscr{C}_c$, and the sheaf of germs of nonvanishing complex valued continuous functions $\mathscr{C}_c^*$. They form an exact sequence of sheaves

$$0 \to \mathscr{Z} \xrightarrow{j} \mathscr{C}_c \xrightarrow{e^{2\pi i}} \mathscr{C}_c^* \to 0.$$

The associated sequence of presheaves of cross-sections was already considered in example 2.7 and it was proved that it is not necessarily exact.

As a further example, we shall consider the sequence of sheaves of germs of the presheaves in the sequence 6.2.

The sheaf of germs $\mathscr{C}^q$ of the presheaf C^q is called the *Alexander–Spanier sheaf of q-dimensional cochains of X*. We remark that, for $G = \mathbf{R}$, $\mathscr{C}^0$ is just the sheaf of germs of real valued functions on X.

We shall prove that the sequence of sheaves

7.2 $$0 \to \mathscr{G} \xrightarrow{\varepsilon} \mathscr{C}^0 \xrightarrow{\delta 0} \mathscr{C}^1 \xrightarrow{\delta 1} \mathscr{C}^2 \xrightarrow{\delta 2} \cdots$$

is exact. Since the sequence of the corresponding presheaves is exact at G and C^0, it follows that the sequence of sheaves

$$0 \to \mathscr{G} \xrightarrow{\varepsilon} \mathscr{C}^0 \xrightarrow{\delta 0} \mathscr{C}^1$$

is also exact. Therefore, we must prove that, if $x \times X$, the sequence of abelian groups

$$\mathscr{C}^0_x \xrightarrow{\delta 0} \mathscr{C}^1_x \xrightarrow{\delta 1} \mathscr{C}^2_x \xrightarrow{\delta 2} \mathscr{C}^3_x \to \cdots$$

is exact.

To do so we define for every $q \geq 0$ a map

$$s^{q+1} : \mathscr{C}^{q+1}_x \to \mathscr{C}^q_x$$

in the following way: If $\not{f} \in \mathscr{C}^{q+1}_x$, let $f \in \text{Hom}(U^{q+2}, G)$ be a representative of the germ $\not{f}$. We define now

$$(s^{q+1}_U f)(x_0, \ldots, x_q) = f(x, x_0, \ldots, x_q).$$

It is easy to check that the map s^{q+1} is well defined and that

$$(s^{q+2} \circ \delta^{q+1} + \delta^q \circ s^{q+1}) \not{f} = \not{f}.$$

Therefore, if $\delta^{q+1}\not{f} = 0$, we have $\delta^q(s^{q+1}\not{f}) = \not{f}$; that is, every element of $\text{Ker}\, \delta^{q+1}$ is an element of $\text{Im}\, \delta^q$. Since it is trivial that $\text{Im}\, \delta^q \subset \text{Ker}\, \delta^{q+1}$, it follows that for every $q \geq 0$, $\text{Im}\, \delta^q = \text{Ker}\, \delta^{q+1}$.

Let $\mathscr{E}$, $\mathscr{F}$ be two sheaves of abelian groups over a space X and let $\varphi : \mathscr{E} \to \mathscr{F}$ be a sheaf homomorphism. For every open set $U \subset X$ there is an induced group homomorphism

$$\phi_U : \Gamma(U, \mathscr{E}) \to \Gamma(U, \mathscr{F}).$$

We may consider $\phi_U \Gamma(U, \mathscr{E})$ as a subgroup of $\Gamma(U, \mathscr{F})$. The factor group

$$G_U = \Gamma(U, \mathscr{F}) / \phi_U \Gamma(U, \mathscr{E})$$

is well defined and it is an abelian group. Let

$$\psi_U : \Gamma(U, \mathscr{F}) \to G_U$$

be the natural projection. Then $G = \{G_U, \varrho^U_V\}$, where ϱ^U_V is the natural restriction map, is a presheaf of abelian groups over X. Let $\mathscr{G}$ be the sheaf of germs of G. The homomorphisms $\{\psi_U\}$ clearly commute with the restriction maps $\{\varrho_V^U\}$ and therefore define a presheaf homomorphism $\psi: \Gamma(\mathscr{F}) \to G$. Then we have a surjective sheaf homomorphism

$$\psi : \mathscr{F} \to \mathscr{G}.$$

Moreover,

$$\mathscr{K}er\, \psi = \mathscr{I}m\, \varphi.$$

The sheaf $\mathscr{G}$ is called the cokernel of φ:

$$\mathscr{G} = \mathscr{C}oker\, \varphi = \mathscr{F}/\mathscr{I}m\, \varphi.$$

In a similar way, we put:

$$\mathscr{C}oim\, \varphi = \mathscr{E}/\mathscr{K}er\, \varphi.$$

Hence, to every sheaf homomorphism $\varphi: \mathscr{E} \to \mathscr{F}$ there are associated two exact sequences of sheaves:

$$0 \to \mathscr{K}er\, \varphi \to \mathscr{E} \to \mathscr{C}oim\, \varphi \to 0,$$

$$0 \to \mathscr{I}m\, \varphi \to \mathscr{F} \to \mathscr{C}oker\, \varphi \to 0.$$

8.1 Flabby sheaves

We have remarked that, if

$$0 \to \mathscr{E} \to \mathscr{F} \to \mathscr{G} \to 0$$

is an exact sequence of sheaves over X and U is any open set in X, the associated sequence of abelian groups

$$0 \to \Gamma(U, \mathscr{E}) \to \Gamma(U, \mathscr{F}) \to \Gamma(U, \mathscr{G}) \to 0$$

is not necessarily exact since the homomorphism $\Gamma(U, \mathscr{F}) \to \Gamma(U, \mathscr{G})$ need not be surjective. Since the exactness of the sequence of cross-sections may provide information on global properties of X, we shall consider some particular sheaves which make that sequence exact.

8.2 DEFINITION *A sheaf $\mathscr{F}$ is said to be flabby if, for every open set $U \subset X$, the map:*

$$\Gamma(X, \mathscr{F}) \to \Gamma(U, \mathscr{F})$$

is surjective.

8.3 *Example* Let $\mathscr{F}$ be any sheaf over X. We denote by $\mathscr{C}^0(X, \mathscr{F})$ the sheaf of germs of the presheaf of arbitrary cross-sections of $\mathscr{F}$; that is, for $U \subset X$, the elements of this presheaf are maps $\sigma: U \to \mathscr{F}$ such that $p \circ \sigma(x) = x$ for every $x \in U$ with no condition of continuity. The sheaf $\mathscr{C}^0(X, \mathscr{F})$ is clearly flabby.

We could prove now that if $0 \to \mathscr{E} \to \mathscr{F} \to \mathscr{G} \to 0$ is an exact sequence of sheaves and $\mathscr{E}$ is flabby, the sequence $0 \to \Gamma(U, \mathscr{E}) \to \Gamma(U, \mathscr{F}) \to \Gamma(U, \mathscr{G}) \to 0$ is exact for every $U \subset X$. Since we shall not need this result, we omit the proof. In fact, if the base space X of $\mathscr{F}$ satisfies some conditions, this result follows as a corollary of a more general theorem.

9.1 Soft sheaves

From now on we shall always assume that the common base space X of our sheaves is a paracompact Hausdorff space. If $\phi: \mathscr{E} \to \mathscr{F}$ is a sheaf homomorphism, we will denote also by $\phi: \Gamma(U, \mathscr{E}) \to \Gamma(U, \mathscr{F})$ the induced homomorphism.

9.2 DEFINITION *A sheaf $\mathscr{F}$ over X is called soft if every cross-section of $\mathscr{F}$ over a closed subset of X extends to a cross-section of $\mathscr{F}$ over X; that is, if for every closed set $K \subset X$, the map $\Gamma(X, \mathscr{F}) \to \Gamma(K, \mathscr{F})$ is surjective.*

It is easy to check that the sheaf $\mathscr{C}^0(X, \mathscr{F})$ in example 8.3 is a soft sheaf.

9.3 THEOREM *Let*

$$0 \to \mathscr{E} \xrightarrow{\phi} \mathscr{F} \xrightarrow{\psi} \mathscr{G} \to 0$$

be an exact sequence of sheaves over X. If $\mathscr{E}$ is soft, the sequence

$$0 \to \Gamma(X, \mathscr{E}) \xrightarrow{\phi} \Gamma(X, \mathscr{F}) \xrightarrow{\psi} \Gamma(X, \mathscr{G}) \to 0$$

is exact.

Proof We have only to prove the exactness at $\Gamma(X, \mathscr{G})$. Let σ be an element of $\Gamma(X, \mathscr{G})$. Since ψ is surjective at sheaf level, σ can be lifted locally to $\Gamma(X, \mathscr{F})$; that is, to every point $x \in X$ there are a neighborhood U of x and a cross-section $\tau \in \Gamma(U, \mathscr{F})$ such that $\psi(\tau) = \sigma | U$.

Since X is paracompact, there exist a locally finite open covering $\{U_i\}_{i \in I}$ of X and cross-sections $\tau_i \in \Gamma(U_i, \mathscr{F})$ such that $\psi(\tau_i) = \sigma | U_i$. Let $\{V_i\}_{i \in I}$ be a covering of X such that, for every i,

$$\bar{V}_i \subset U_i.$$

Denote by (J, τ) a pair consisting of a subset $J \subset I$ and an element $\tau \in \Gamma(\bar{V}_J, \mathscr{F})$ where $\bar{V}_J = \bigcup_{i \in J} \bar{V}_i$ and $\psi(\tau) = \sigma | \bar{V}_J$. The set of all pairs (J, τ) is inductively ordered. Let (J, τ) be a maximal element. We must show that $J = I$.

Assume not and let $i \in I - J$. There exist elements $\tau \in \Gamma(\bar{V}_J, \mathscr{F})$ and $\tau_i \in \Gamma(\bar{V}_i, \mathscr{F})$ such that $\psi(\tau) = \sigma | \bar{V}_J$ and $\psi(\tau_i) = \sigma | \bar{V}_i$. Now, on $\bar{V}_J \cap \bar{V}_i$ we have $\psi(\tau - \tau_i) = 0$ or $\tau - \tau_i \in \operatorname{Ker} \psi$. Because of the exactness, there exists an element $\eta \in \Gamma(\bar{V}_J \cap \bar{V}_i, \mathscr{E})$ such that $\phi(\eta) = \tau - \tau_i$. Since $\mathscr{E}$ is soft and $\bar{V}_J \cap \bar{V}_i$ is closed, η can be extended to X; that is, there is an element $\hat{\eta} \in \Gamma(X, \mathscr{E})$ such that $\hat{\eta} | \bar{V}_J \cap \bar{V}_i = \eta$. Now $\tau_i + \psi(\hat{\eta}) \in \Gamma(\bar{V}_i, \mathscr{F})$ and $\tau = \tau_i + \phi(\hat{\eta})$ on $\bar{V}_J \cap \bar{V}_i$. Thus, there is an element $\xi \in \Gamma(\bar{V}_J \cap \bar{V}_i, \mathscr{F})$ such that $\xi | \bar{V}_J = \tau$ and $\xi | \bar{V}_i = \tau_i + \phi(\hat{\eta})$. Now, since $\psi(\phi(\hat{\eta})) = 0$, we have $\psi(\xi) = \sigma | \bar{V}_J \cap \bar{V}_i$ and this contradicts the hypothesis that (J, τ) is maximal. ○

9.4 THEOREM *Every flabby sheaf is soft.*

Proof It is enough to prove that every cross-section of a flabby sheaf $\mathscr{F}$ over a closed set $K \subset X$ can be extended to an open neighborhood of K.

Let σ be a cross-section of $\mathscr{F}$ over K. Let $\{U_i\}$ be a locally finite open covering of X and let $\sigma_i \in \Gamma(U_i, \mathscr{F})$ such that $\sigma_i = \sigma | U_i \cap K$.

Let $\mathscr{V} = \{V_i\}$ be a covering of X such that, for every i, $\bar{V}_i \subset U_i$. Let U be the set of all $x \in X$ such that, if $x \in \bar{V}_i \cap \bar{V}_j$, then $\sigma_i(x) = \sigma_j(x)$. It is easy to show that there exists a cross-section $\hat{\sigma}$ over U such that $\hat{\sigma} | U \cap \bar{V}_i = \sigma_i$ and $\hat{\sigma} | K = \sigma$.

It remains to show that U is an open neighborhood of K. Every point $x \in K$ has an open neighborhood $U(x)$ which meets only a finite number $\bar{V}_{i_1}, \ldots, \bar{V}_{i_p}$ of elements of $\mathscr{V}$ and we can assume that $U(x)$ is contained in the corresponding sets $U_{i_1}, \ldots, U_{i_p}$. Then we have $\sigma_{i_1}(x) = \cdots = \sigma_{i_p}(x)$. Thus, we can have $\sigma_{i_1} | U(x) = \cdots = \sigma_{i_p} | U(x)$. Hence, $U(x) \subset U$. ○

9.5 COROLLARY *If $0 \to \mathscr{E} \to \mathscr{F} \to \mathscr{G} \to 0$ is an exact sequence of sheaves and $\mathscr{E}$ is flabby, the sequence*

$$0 \to \Gamma(U, \mathscr{E}) \to \Gamma(U, \mathscr{F}) \to \Gamma(U, \mathscr{G}) \to 0$$

is exact for every open set $U \subset X$.

The proof is a straightforward consequence of theorems 9.3, 9.4. ○

It follows from the above corollary that the sequence of presheaves

$$0 \to \Gamma(\mathscr{E}) \to \Gamma(\mathscr{F}) \to \Gamma(\mathscr{G}) \to 0$$

is also exact.

9.6 THEOREM *Let $0 \to \mathscr{E} \to \mathscr{F} \xrightarrow{\psi} \mathscr{G} \to 0$ be an exact sequence of sheaves. If $\mathscr{E}$ and $\mathscr{F}$ are soft, then $\mathscr{G}$ is soft.*

Proof We must show that every cross-section σ of $\mathscr{G}$ over a closed set $K \subset X$ can be extended to a cross-section $\hat{\sigma}$ of $\mathscr{G}$ over X. Since $\mathscr{E}$ is soft, this follows at once from theorem 9.3. In fact, σ is the image under ψ of a cross-section τ of $\mathscr{F}$ over K which extends to a cross-section $\hat{\tau}$ over X since $\mathscr{F}$ is soft. It follows that $\psi(\hat{\tau})$ is a cross-section of $\mathscr{G}$ over X such that $\psi(\hat{\tau})\,|\,K = \sigma$. ○

9.7 THEOREM *Let*

$$0 \to \mathscr{F}^0 \xrightarrow{\phi^0} \mathscr{F}^1 \xrightarrow{\phi^1} \mathscr{F}^2 \to \cdots$$

be an exact sequence of soft sheaves over X. The sequence of abelian groups

$$0 \to \Gamma(X, \mathscr{F}^0) \xrightarrow{\phi^0} \Gamma(X, \mathscr{F}^1) \xrightarrow{\phi^1} \Gamma(X, \mathscr{F}^2) \to \cdots$$

is also exact.

Proof For $p = 0, 1, \ldots$ we put $\mathscr{Z}^p = \mathscr{K}er\, \phi^p$; for $p > 0$ we also have $\mathscr{Z}^p = \mathscr{I}m\, \phi^{p-1}$. The exactness of the long sequence of sheaves is equivalent to the exactness of the short sequences

$$0 \to \mathscr{Z}^p \to \mathscr{F}^p \to \mathscr{Z}^{p+1} \to 0.$$

Since $\mathscr{Z}^0 = \mathscr{K}er\, \phi^0 = 0$ and $\mathscr{F}^0$ are soft sheaves, it follows by theorem 7.6 that $\mathscr{Z}^1$ is soft. Hence, for every p, $\mathscr{Z}^p$ is a soft sheaf. Then by theorem 9.3 the sequences

$$0 \to \Gamma(X, \mathscr{Z}^p) \to \Gamma(X, \mathscr{F}^p) \to \Gamma(X, \mathscr{Z}^{p+1}) \to 0$$

are exact for every p. This is equivalent to

$$\operatorname{Ker}\big(\Gamma(X, \mathscr{F}^p) \xrightarrow{\phi^p} \Gamma(X, \mathscr{F}^{p+1})\big) = \operatorname{Im}\big(\Gamma(X, \mathscr{F}^{p-1}) \xrightarrow{\phi^{p-1}} \Gamma(X, \mathscr{F}^p)\big)$$

for every p. ○

It is clear that theorem 9.7 is still valid if X is replaced by any open set $U \subset X$.

10.1 Fine sheaves

We shall consider a third class of sheaves which we shall use later, but first we need some more properties of soft sheaves.

10.2 THEOREM *Let $\mathscr{A}$ be a sheaf of commutative rings with unity over X. Then $\mathscr{A}$ is soft if and only if every point $x \in X$ has a neighborhood U such that given two disjoint closed sets $K, K' \subset U$ there exists a cross-section $\sigma \in \Gamma(U, \mathscr{A})$ such that $\sigma | K = 1$ and $\sigma | K' = 0$.*

Proof It is enough to prove that, if K is a closed subset of X, U is an open neighborhood of K and σ is a cross-section of $\mathscr{A}$ over K, then σ can be extended to a cross-section which is zero on the boundary ∂U of U. Since $\mathscr{A}$ is soft there exists a cross-section τ which is 1 over K and 0 on ∂U. Then the cross-section $\hat{\sigma}(x) = \tau(x)\,\sigma(x)$ has the property that $\hat{\sigma} | K = \sigma$ and $\hat{\sigma}(x) = 0$ if $x \in \partial U$.

The sufficiency is trivial. ○

10.3 COROLLARY *Every $\mathscr{A}$-Module over a soft sheaf of rings is soft.*

10.4 DEFINITION *A sheaf $\mathscr{F}$ of abelian groups over X is said to be fine if the sheaf of germs of homomorphisms $\mathscr{H}om_{Z}(\mathscr{F}, \mathscr{F})$ is soft.*

It follows from theorem 10.2 that a sheaf $\mathscr{F}$ of abelian groups over X is fine if and only if, given two disjoint closed subsets A, B of X, there exists a sheaf homomorphism $\phi: \mathscr{F} \to \mathscr{F}$ such that $\phi: \mathscr{F} | A \to \mathscr{F} | A$ is the identity homomorphism and $\phi: \mathscr{F} | B \to \mathscr{F} | B$ is the zero homomorphism.

Then if $\mathscr{F}$ is a fine sheaf we can define partitions of unity of the sheaf $\mathscr{F}$. If $\{U_i\}$ is a locally finite open covering of X, a collection of sheaf homomorphisms $\{\phi_i\}: \mathscr{F} \to \mathscr{F}$ is called a partition of unity of $\mathscr{F}$ subordinate to the covering $\{U_i\}$ if:

i) ϕ_i is the zero homomorphism on a neighborhood of the complement of U_i,

ii) $\sum_i \phi_i =$ identity homomorphism.

On the other hand, the existence of a partition of unity for a sheaf $\mathscr{F}$ clearly implies that $\mathscr{F}$ is fine.

10.5 THEOREM *Every fine sheaf is soft.*

Proof It is similar to that of theorem 10.2. Let $\mathscr{F}$ be a fine sheaf over X, let K be a closed subset of X and let σ be a cross-section of $\mathscr{F}$ over K. To every point $x \in X$ there are a neighborhood U and an element $\tau \in \Gamma(U, \mathscr{F})$ such that $\sigma = \tau$ on $U \cap K$. We can assume that these open sets belong to a locally finite open covering $\{U_i\}$ of X. Thus, to every U_i there is an element $\tau_i \in \Gamma(U_i, \mathscr{F})$ such that $\tau_i | U_i \cap K = \sigma$ if $U_i \cap K \neq \emptyset$ and $\tau_i = 0$ if

$U_i \cap K = \emptyset$. Let $\{\varphi_i\}$ be a partition of unity of $\mathscr{F}$ subordinate to the covering $\{U_i\}$. It is easy to check that the cross-sections

$$\eta_i(x) = \begin{cases} \phi_i(\tau_i)(x) & \text{if} \quad x \in U_i \\ 0 & \text{if} \quad x \in X \backslash U_i \end{cases}$$

are elements of $\Gamma(X, \mathscr{F})$ and that $\eta = \sum_i \eta_i \in \Gamma(X, \mathscr{F})$ has the property that $\eta | K = \sigma$. ○

10.6 Example Let M be a differentiable manifold. To every locally finite open covering $\{U_i\}$ of M, there is a subordinate C^∞ partition of unity $\{f_i\}$. We consider now the sheaf of germs of differentiable (or continuous) functions on M. On this sheaf the multiplication of functions is a sheaf homomorphism. In particular, the multiplication by the functions $\{f_i\}$ gives a partition of unity of the sheaf subordinate to the covering $\{U_i\}$.

Thus, the sheaf of germs of differentiable functions on a differentiable manifold M and the sheaf of germs of continuous functions on M are fine sheaves. They are also soft by theorem 10.5.

We remark that the sheaf of germs of holomorphic functions on a complex manifold M is not a soft sheaf.

It follows from corollary 10.3 that sheaves of germs of differential forms either on a differentiable manifold or on a complex manifold are always soft sheaves. Actually, they are fine sheaves, as it is easy to verify.

11.1 Problems

1 (3.1) Let $\mathscr{E}$ be a subsheaf of a sheaf $\mathscr{F}$. Prove that the quotient sheaf $\mathscr{F}/\mathscr{E}$ is well defined and that the sequence

$$0 \to \mathscr{E} \to \mathscr{F} \to \mathscr{F}/\mathscr{E} \to 0$$

is exact.

2 (3.1) Let $\mathscr{E}, \mathscr{F}$ be sheaves of abelian groups over a topological space X. Define the tensor product $\mathscr{E} \otimes \mathscr{F}$ by $(\mathscr{E} \otimes \mathscr{F})_x = \mathscr{E}_x \otimes \mathscr{F}_x$. Show that $\mathscr{E} \otimes \mathscr{F}$ satisfies the following universal property: if $\mathscr{G}$ is any sheaf of abelian groups over X and $\phi : \mathscr{E} \underset{X}{\times} \mathscr{F} \to \mathscr{G}$ is a continuous map which commutes with the projections and is bilinear on every stalk, there exists a unique sheaf-homomorphism $\Psi : \mathscr{E} \otimes \mathscr{F} \to \mathscr{G}$ such that the following diagram

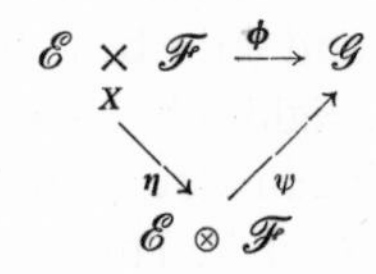

commutes. $\eta : \mathscr{E} \underset{X}{\times} \mathscr{F} \to \mathscr{E} \otimes \mathscr{F}$ is the map taking $(e, f) \in \mathscr{E}_x \times \mathscr{F}_x$ into $e \otimes f \in (\mathscr{E} \otimes \mathscr{F})_x$.

3 (4.1) Let $\mathscr{E}$, $\mathscr{F}$ be two *A-Modules* over X. Show that the sheaf $\mathscr{H}om_{\mathscr{A}}(\mathscr{E},\mathscr{F})$ of germs of homomorphisms from $\mathscr{E}$ to $\mathscr{F}$, in general, is not an *A-Module*.

4 (4.1) Let T be the reals mod. 1 and let X be any space. Let $\mathscr{F}$ be the sheaf of germs of continuous maps $X \to \mathbf{R}$ and let τ be the sheaf of germs of continuous maps $X \to T$. Show that there is an exact sequence of sheaves over X:

$$0 \to \mathscr{Z} \to \mathscr{F} \to \tau \to 0.$$

5 (4.1) A sheaf $p : \mathscr{E} \to X$ is said to be *locally constant* if every $x \in X$ has a neighborhood $U(x)$ such that $(p^{-1}(U(x)), p \mid p^{-1}(U(x)), U(x))$ is (isomorphic to) a constant sheaf. Show that not every locally constant sheaf derives from a locally constant presheaf as defined in example 1.5. Also show that every constant sheaf may be defined by some locally constant presheaf.

6 (8.1) If $\mathscr{F}$ is a sheaf over X such that $\mathscr{F} \mid U$ is flabby for every sufficiently small open set $U \subset X$, show that $\mathscr{F}$ is flabby.

CHAPTER 8

Cohomology with coefficients in a sheaf

1.1 Resolutions

LET $\mathscr{E}$ BE a sheaf of abelian groups over X. An exact sequence of sheaves of the form:

1.2 $$0 \to \mathscr{E} \xrightarrow{\varepsilon} \mathscr{F}^0 \xrightarrow{\phi_0} \mathscr{F}^1 \xrightarrow{\phi_1} \mathscr{F}^2 \to \cdots$$

is called a *cohomological resolution* of the sheaf $\mathscr{E}$. If for every $n = 0, 1, \ldots$, the sheaf $\mathscr{F}^n$ is flabby, soft, or fine, the sequence 1.2 is called a flabby, soft, or fine resolution of $\mathscr{E}$ respectively.

As an example, the sequence 7.7.2 of Alexander–Spanier sheaves of cochains is a cohomological resolution of the sheaf $\mathscr{G}$.

To the resolution 1.2 we can associate a sequence of global cross-sections

1.3 $$0 \to \Gamma(X, \mathscr{E}) \xrightarrow{\varepsilon} \Gamma(X, \mathscr{F}^0) \xrightarrow{\phi_0} \Gamma(X, \mathscr{F}^1) \xrightarrow{\phi_1} \Gamma(X, \mathscr{F}^2) \to \cdots$$

which in general is not exact. Of course, if 1.2 is a soft resolution of $\mathscr{E}$ and $\mathscr{E}$ is a soft sheaf, then the sequence 1.3 is exact by theorem 7.9.7.

In any case, we have that $\operatorname{Im} \phi_n \subset \operatorname{Ker} \phi_{n+1}$ for $n = 0, 1, \ldots$.

1.4 THEOREM *To every sheaf $\mathscr{F}$ over X there is a soft resolution of $\mathscr{F}$:*

1.5 $$0 \to \mathscr{F} \xrightarrow{\varepsilon} \mathscr{C}^0(X, \mathscr{F}) \xrightarrow{d_0} \mathscr{C}^1(X, \mathscr{F}) \xrightarrow{d_1} \mathscr{C}^2(X, \mathscr{F}) \to \cdots.$$

Proof Let $\mathscr{C}^0(X, \mathscr{F})$ be the sheaf of germs of arbitrary cross-sections of $\mathscr{F}$ over X defined in example 7.8.3. This sheaf is soft. We put now $\mathscr{Z}^0(X, \mathscr{F}) = \mathscr{F}$ and $\mathscr{Z}^1(X, \mathscr{F}) = \mathscr{C}^0(X, \mathscr{F})/j_0\mathscr{F}$ where $j_0\colon \mathscr{F} \to \mathscr{C}^0(X, \mathscr{F})$ is the natural embedding; that is, $\mathscr{Z}^1(X, \mathscr{F}) = \mathscr{C}\mathit{oker}\, j_0$. Then we have the exact sequence

$$0 \to \mathscr{Z}^0(X, \mathscr{F}) \xrightarrow{j_0} \mathscr{C}^0(X, \mathscr{F}) \xrightarrow{p_0} \mathscr{Z}^1(X, \mathscr{F}) \to 0$$

where $p_0\colon \mathscr{C}^0(X, \mathscr{F}) \to \mathscr{Z}^1(X, \mathscr{F}) = \mathscr{C}\mathit{oker}\, j_0$ is the natural projection. Now by induction on n, assume we have defined $\mathscr{Z}^n(X, \mathscr{F})$ and let $\mathscr{C}^n(X, \mathscr{F}) = \mathscr{C}^0(X, \mathscr{Z}^n(X, \mathscr{F}))$. Let $j_n\colon \mathscr{Z}^n(X, \mathscr{F}) \to \mathscr{C}^n(X, \mathscr{F})$ be the natural embedding and define $\mathscr{Z}^{n+1}(X, \mathscr{F}) = \mathscr{C}\mathit{oker}\, j_n$. Then for $n = 0, 1, \ldots$ we have exact sequences

$$0 \to \mathscr{Z}^n(X, \mathscr{F}) \xrightarrow{j_n} \mathscr{C}^n(X, \mathscr{F}) \xrightarrow{p_n} \mathscr{Z}^{n+1}(X, \mathscr{F}) \to 0.$$

Putting $\mathscr{Z}^n(X, \mathscr{F}) = \mathscr{Z}^n$ and $\mathscr{C}^n(X, \mathscr{F}) = \mathscr{C}^n$ and patching all these sequences together we get a long sequence

$$0 \to \mathscr{Z}^0 \xrightarrow{j_0} \mathscr{C}^0 \xrightarrow{p_0} \mathscr{Z}^1 \xrightarrow{j_1} \mathscr{C}^1 \xrightarrow{p_1} \mathscr{Z}^2 \xrightarrow{j_2} \mathscr{C}^2 \to \cdots.$$

Letting $\varepsilon = j_0$ and $d_n = j_{n+1} \circ p_n$, we obtain the sequence

$$0 \to \mathscr{F} \xrightarrow{\varepsilon} \mathscr{C}^0(X, \mathscr{F}) \xrightarrow{d_0} \mathscr{C}^1(X, \mathscr{F}) \xrightarrow{d_1} \mathscr{C}^2(X, \mathscr{F}) \to \cdots,$$

which is exact by construction. Since for every $n = 0, 1, \ldots, \mathscr{C}^n(X, \mathscr{F})$ is a soft sheaf, this exact sequence is a soft resolution of $\mathscr{F}$. ○

The sequence 1.5 is called the *canonical resolution* of the sheaf $\mathscr{F}$.

1.6 THEOREM *Let $\phi : \mathscr{E} \to \mathscr{F}$ be a sheaf homomorphism. Then ϕ induces a commutative diagram*

$$\begin{array}{ccccccccc}
0 \to \mathscr{E} & \xrightarrow{\varepsilon} & \mathscr{C}^0(X, \mathscr{E}) & \xrightarrow{d_0} & \mathscr{C}^1(X, \mathscr{E}) & \xrightarrow{d_1} & \mathscr{C}^2(X, \mathscr{E}) & \to \cdots \\
\downarrow \phi & & \downarrow \phi_0 & & \downarrow \phi_1 & & \downarrow \phi_2 & \\
0 \to \mathscr{F} & \xrightarrow{\varepsilon'} & \mathscr{C}^0(X, \mathscr{F}) & \xrightarrow{d'_0} & \mathscr{C}^1(X, \mathscr{F}) & \xrightarrow{d'_1} & \mathscr{C}^2(X, \mathscr{F}) & \to \cdots
\end{array}$$

of canonical resolutions.

Proof For every open set $U \subset X$, the sheaf homomorphism ϕ induces a homomorphism of groups of arbitrary cross-sections over U, ϕ_{0U}: $C(U, \mathscr{E}) \to C(U, \mathscr{F})$, thus, a presheaf homomorphism which, in turn, defines a sheaf homomorphism ϕ_0: $\mathscr{C}^0(X, \mathscr{E}) \to \mathscr{C}^0(X, \mathscr{F})$. There is also a sheaf homomorphism $\bar{\phi}_1$: $\mathscr{C}oker\, j_0 \to \mathscr{C}oker\, j'_0$. Hence we have a diagram

$$\begin{array}{ccccccc}
0 \to \mathscr{E} & \xrightarrow{j_0} & \mathscr{C}^0(X, \mathscr{E}) & \xrightarrow{p_0} & \mathscr{Z}^1(X, \mathscr{E}) & \to 0 \\
\downarrow \phi & & \downarrow \phi_0 & & \downarrow \bar{\phi}_1 & \\
0 \to \mathscr{F} & \xrightarrow{j'_0} & \mathscr{C}^0(X, \mathscr{F}) & \xrightarrow{p'_0} & \mathscr{Z}^1(X, \mathscr{F}) & \to 0
\end{array}$$

with exact rows which is clearly commutative. Then, by induction, for every $n = 0, 1, \ldots,$ we have commutative diagrams

$$\begin{array}{ccccccc}
0 \to \mathscr{Z}^n(X, \mathscr{E}) & \to & \mathscr{C}^n(X, \mathscr{E}) & \to & \mathscr{Z}^{n+1}(X, \mathscr{E}) & \to 0 \\
\downarrow \bar{\phi}_n & & \downarrow \phi_n & & \downarrow \bar{\phi}_{n+1} & \\
0 \to \mathscr{Z}^n(X, \mathscr{F}) & \to & \mathscr{C}^n(X, \mathscr{F}) & \to & \mathscr{Z}^{n+1}(X, \mathscr{F}) & \to 0.
\end{array}$$

The proof of the theorem now follows easily by putting together all these diagrams. ○

We remark that, if ϕ is the identity, then, for every n, ϕ_n is also the identity.

1.7 COROLLARY *If*

$$\begin{array}{ccccccccc} 0 \to & \mathscr{E}' & \xrightarrow{j} & \mathscr{E} & \xrightarrow{p} & \mathscr{E}'' & \to 0 \\ & \downarrow \phi & & \downarrow \psi & & \downarrow \eta & \\ 0 \to & \mathscr{F}' & \xrightarrow{j'} & \mathscr{F} & \xrightarrow{p'} & \mathscr{F}'' & \to 0 \end{array}$$

is a commutative diagram of sheaves with exact rows, then for every $n = 0, 1, \cdots$ *the diagram*

$$\begin{array}{ccccccccc} 0 \to & \mathscr{C}^n(X, \mathscr{E}') & \xrightarrow{j_n} & \mathscr{C}^n(X, \mathscr{E}) & \xrightarrow{p_n} & \mathscr{C}^n(X, \mathscr{E}'') & \to 0 \\ & \downarrow \phi_n & & \downarrow \psi_n & & \downarrow \eta_n & \\ 0 \to & \mathscr{C}^n(X, \mathscr{F}') & \xrightarrow{j'_n} & \mathscr{C}^n(X, \mathscr{F}) & \xrightarrow{p'_n} & \mathscr{C}^n(X, \mathscr{F}'') & \to 0 \end{array}$$

has exact rows and commutes.

Proof By induction on n. ○

2.1 Cohomology groups

Let $\mathscr{F}$ be a sheaf of abelian groups over X. Let

$$0 \to \mathscr{F} \to \mathscr{C}^0(X, \mathscr{F}) \to \mathscr{C}^1(X, \mathscr{F}) \to \mathscr{C}^2(X, \mathscr{F}) \to \cdots$$

be the canonical resolution of $\mathscr{F}$ and let

$$0 \to \Gamma(X, \mathscr{F}) \xrightarrow{\varepsilon} \Gamma(X, \mathscr{C}^0) \xrightarrow{d_0} \Gamma(X, \mathscr{C}^1) \xrightarrow{d_1} \Gamma(X, \mathscr{C}^2) \to \cdots$$

be the associated sequence of global cross-sections.

2.2 DEFINITION *For* $n \geq 1$, *the abelian group*

$$H^n(X, \mathscr{F}) = \frac{\operatorname{Ker}\left(\Gamma(X, \mathscr{C}^n) \xrightarrow{d_n} \Gamma(X, \mathscr{C}^{n+1})\right)}{\operatorname{Im}\left(\Gamma(X, \mathscr{C}^{n-1}) \xrightarrow{d_{n-1}} \Gamma(X, \mathscr{C}^n)\right)}$$

is called the n-th cohomology group of X with coefficients in the sheaf $\mathscr{F}$. *For* $n = 0$ *we put*

$$H^0(X, \mathscr{F}) = \operatorname{Ker} d_0.$$

With the above definition we associate to every sheaf $\mathscr{F}$ over X a sequence $H^0(X, \mathscr{F})$, $H^1(X, \mathscr{F})$, ... of cohomology groups. We shall now show some properties of these groups and then prove that they are characterized by these properties.

2.3 THEOREM *The cohomology groups of X with coefficients in a sheaf satisfy the following "functorial properties":*

i) *if* $\phi: \mathscr{E} \to \mathscr{F}$ *is a sheaf homomorphism, then for every* $n = 0, 1, \ldots$ *there is an induced homomorphism* $\phi_n: H^n(X, \mathscr{E}) \to H^n(X, \mathscr{F})$;

ii) *if* ϕ *is the identity, then for every* $n = 0, 1, \ldots, \phi_n$ *is the identity;*

iii) *if* $\eta: \mathscr{E} \to \mathscr{G}$ *is the composition of the sheaf homomorphisms* $\mathscr{E} \xrightarrow{\phi} \mathscr{F} \xrightarrow{\psi} \mathscr{G}$, *then, for every* $n = 0, 1, \ldots, \eta_n = \psi_n \circ \phi_n$.

Proof It is a trivial application of theorem 1.6 and definitions. ○

2.4 THEOREM *For every sheaf* $\mathscr{F}$ *over X, the cohomology group* $H^0(X, \mathscr{F})$ *is isomorphic to* $\Gamma(X, \mathscr{F})$.

Proof Since the sequence

$$0 \to \Gamma(X, \mathscr{F}) \xrightarrow{\varepsilon} \Gamma(X, \mathscr{C}^0) \xrightarrow{d_0} \Gamma(X, \mathscr{C}^1)$$

is exact, the group $\Gamma(X, \mathscr{F})$ may be identified with $\operatorname{Ker} d_0 = H^0(X, \mathscr{F})$. ○

2.5 *Remark* It is easy to verify that if $\phi: \mathscr{E} \to \mathscr{F}$ is a sheaf homomorphism, the induced ϕ_0: $H^0(X, \mathscr{E}) \to H^0(X, \mathscr{F})$ is exactly the homomorphism $\phi: \Gamma(X, \mathscr{E}) \to \Gamma(X, \mathscr{F})$.

2.6 THEOREM *To every exact sequence*

$$0 \to \mathscr{F}' \xrightarrow{j} \mathscr{F} \xrightarrow{p} \mathscr{F}'' \to 0$$

of sheaves over X there corresponds an exact sequence of cohomology groups

$$0 \to H^0(X, \mathscr{F}') \xrightarrow{j_0} H^0(X, \mathscr{F}) \xrightarrow{p_0} H^0(X, \mathscr{F}'') \xrightarrow{\delta_0} H^1(X, \mathscr{F}') \to \cdots$$

$$\cdots \to H^n(X, \mathscr{F}') \xrightarrow{j_n} H^n(X, \mathscr{F}) \xrightarrow{p_n} H^n(X, \mathscr{F}'') \xrightarrow{\delta_n} H^{n+1}(X, \mathscr{F}') \to \cdots.$$

Proof To the given exact sequence is associated a diagram

$$\begin{array}{ccccccccc}
 & & \vdots & & \vdots & & \vdots & & \\
 & & \downarrow & & \downarrow & & \downarrow & & \\
0 & \to & \Gamma(X, \mathscr{C}^{n-1}(\mathscr{F}')) & \xrightarrow{j_{n-1}} & \Gamma(X, \mathscr{C}^{n-1}(\mathscr{F})) & \xrightarrow{p_{n-1}} & \Gamma(X, \mathscr{C}^{n-1}(\mathscr{F}'')) & \to & 0 \\
 & & \downarrow d'_{n-1} & & \downarrow d_{n-1} & & \downarrow d''_{n-1} & & \\
0 & \to & \Gamma(X, \mathscr{C}^{n}(\mathscr{F}')) & \xrightarrow{j_{n}} & \Gamma(X, \mathscr{C}^{n}(\mathscr{F})) & \xrightarrow{p_{n}} & \Gamma(X, \mathscr{C}^{n}(\mathscr{F}'')) & \to & 0 \\
 & & \downarrow d'_{n} & & \downarrow d_{n} & & \downarrow d''_{n} & & \\
0 & \to & \Gamma(X, \mathscr{C}^{n+1}(\mathscr{F}')) & \xrightarrow{j_{n+1}} & \Gamma(X, \mathscr{C}^{n+1}(\mathscr{F})) & \xrightarrow{p_{n+1}} & \Gamma(X, \mathscr{C}^{n+1}(\mathscr{F}'')) & \to & 0 \\
 & & \downarrow d'_{n+1} & & \downarrow d_{n+1} & & \downarrow d''_{n+1} & & \\
0 & \to & \Gamma(X, \mathscr{C}^{n+2}(\mathscr{F}')) & \xrightarrow{j_{n+2}} & \Gamma(X, \mathscr{C}^{n+2}(\mathscr{F})) & \xrightarrow{p_{n+2}} & \Gamma(X, \mathscr{C}^{n+2}(\mathscr{F}'')) & \to & 0 \\
 & & \downarrow & & \downarrow & & \downarrow & & \\
 & & \vdots & & \vdots & & \vdots & &
\end{array}$$

In this infinite diagram, every row is exact by theorem 7.9.3 since every sheaf is soft. We now define the homomorphism δ_n: $H^n(X, \mathscr{F}'') \to H^{n+1}(X, \mathscr{F}')$, that is,

$$\delta_n : \operatorname{Ker} d''_n / \operatorname{Im} d''_{n-1} \to \operatorname{Ker} d'_{n+1} / \operatorname{Im} d'_n$$

as follows:

Let $h'' \in \operatorname{Ker} d''_n / \operatorname{Im} d''_{n-1}$; a representative of h'' is an element $a'' \in \operatorname{Ker} d''_n$. Since p_n is surjective, there exists an element $a \in \Gamma(X, \mathscr{C}^n(\mathscr{F}))$ such that $p_n a = a''$. We now have that $p_{n+1} \circ d_n a = d''_n \circ p_n a = d''_n a'' = 0$. By the exactness of the $(n+1)$-th row, since $d_n a \in \operatorname{Ker} p_{n+1}$, it follows that there exists an element $a' \in \Gamma(X, \mathscr{C}^{n+1}(\mathscr{F}'))$ such that $j_{n+1} a' = d_n a$. We now consider $d'_{n+1} a'$. We have $j_{n+2} \circ d'_{n+1} a' = d_{n+1} \circ j_{n+1} a' = d_{n+1} \circ d_n a = 0$. Since j_{n+2} is injective we have $d'_{n+1} a' = 0$. Thus, $a' \in \operatorname{Ker} d'_{n+1}$. We wish to define $\delta_n h'' =$ the class of a'. The definition will be consistent if we show that a' is an element of $\operatorname{Ker} d'_{n+1} / \operatorname{Im} d'_n$ which does not depend on the choices of a'', a, and a'.

Let $\bar{a}''$ be another representative of h'' and let $\bar{a}$, $\bar{a}'$ be other choices. Then we have $a'' - \bar{a}'' = d''_{n-1} c''$ for some c'', and since p_{n-1} is surjective there is a $c \in \Gamma(X, \mathscr{C}^{n-1}(\mathscr{F}))$ such that $c'' = p_{n-1} c$. We now have

$$p_n(a - \bar{a} - d_{n-1} c) = 0,$$

and, by the exactness, there is an element $c' \in \Gamma(X, \mathscr{C}^n(\mathscr{F}'))$ such that

$$a - \bar{a} - d_{n-1}c = j_n c'.$$

Then $j_{n+1} \circ d_n' c' = d_n \circ j_n c' = d_n a - d_n \bar{a} = j_{n+1} a' - j_{n+1} \bar{a}' = j_{n+1}(a' - \bar{a}')$. Since j_{n+1} is injective, it follows that $a' - \bar{a}' = d_n' c'$. Thus, $a' - \bar{a}' \in \operatorname{Im} d_n'$, which proves that δ_n is well defined.

It remains to show that the homomorphisms δ_n make the sequence

$$\cdots \to H^n(X, \mathscr{F}) \xrightarrow{p_n} H^n(X, \mathscr{F}'') \xrightarrow{\delta_n} H^{n+1}(X, \mathscr{F}') \xrightarrow{j_{n+1}} \cdots$$

exact. This is a simple verification which we leave to the reader. ○

2.7 THEOREM *Let*

$$\begin{array}{ccccccccc} 0 & \to & \mathscr{E}' & \to & \mathscr{E} & \to & \mathscr{E}'' & \to & 0 \\ & & \downarrow & & \downarrow & & \downarrow & & \\ 0 & \to & \mathscr{F}' & \to & \mathscr{F} & \to & \mathscr{F}'' & \to & 0 \end{array}$$

be a commutative diagram of sheaves with exact rows. Then for every $n = 0, 1, \cdots$, *the diagram*

$$\begin{array}{ccc} H^n(X, \mathscr{E}'') & \xrightarrow{\delta_n} & H^{n+1}(X, \mathscr{E}') \\ \downarrow & & \downarrow \\ H^n(X, \mathscr{F}'') & \xrightarrow{\delta_n'} & H^{n+1}(X, \mathscr{F}') \end{array}$$

commutes.

Proof It follows at once from corollary 1.7. ○

2.8 THEOREM *If* $\mathscr{F}$ *is a soft sheaf, then* $H^n(X, \mathscr{F}) = 0$ *for* $n \geq 1$.

Proof We consider the sequence

$$0 \to \Gamma(X, \mathscr{F}) \xrightarrow{\varepsilon} \Gamma(X, \mathscr{C}^0) \xrightarrow{d_0} \Gamma(X, \mathscr{C}^1) \xrightarrow{d_1} \Gamma(X, \mathscr{C}^2) \to \cdots.$$

If $\mathscr{F}$ is soft, then every sheaf is soft; hence, by theorem 7.9.7 the sequence is exact. This means that, for $n \geq 1$, $\operatorname{Ker} d_n = \operatorname{Im} d_{n-1}$ or $H^n(X, \mathscr{F}) = 0$. ○

We shall show now that for every sheaf of abelian groups $\mathscr{F}$ over X the associated cohomology groups $H^n(X, \mathscr{F})$ are unique up to isomorphism. In other words, we shall prove that, if we have two sequences of abelian groups $\{H^n(X, \mathscr{F})\}$ and $\{\tilde{H}^n(X, \mathscr{F})\}$ satisfying the conditions of theorems 2.3, 2.4, 2.6, 2.7, and 2.8, then for every $n = 0, 1, \ldots$ there is an isomorphism

$\alpha_n: H^n(X, \mathscr{F}) \to \tilde{H}^n(X, \mathscr{F})$. Moreover, this family of isomorphisms satisfies all the necessary compatibility conditions.

2.9 THEOREM *The cohomology groups $H^n(X, \mathscr{F})$ are unique up to isomorphism.*

Proof We shall not give every detail of the proof, which is rather technical, but only some hints. With the above notation, let $\{H^n(X, \mathscr{F})\}$ and $\{\tilde{H}^n(X, \mathscr{F})\}$ be two sequences of abelian groups both satisfying the conditions of said theorems.

First of all, by theorem 2.4 we have an isomorphism

$$\alpha_0 : H^0(X, \mathscr{F}) \to \tilde{H}^0(X, \mathscr{F}),$$

since both these groups are isomorphic to $\Gamma(X, \mathscr{F})$.

We consider now the exact sequence of sheaves

$$0 \to \mathscr{F} \xrightarrow{j_0} \mathscr{C}^0(X, \mathscr{F}) \xrightarrow{p_0} \mathscr{Z}^1(X, \mathscr{F}) \to 0$$

where $\mathscr{Z}^1(X, \mathscr{F}) = \mathscr{C}oker\, j_0$ as in theorem 1.4. By theorem 2.6, we then have two exact sequences of cohomology groups

$$\begin{array}{ccccccccc} 0 \to & \Gamma(X, \mathscr{F}) & \to & \Gamma(X, \mathscr{C}^0) & \to & \Gamma(X, \mathscr{Z}^1) & \to & H^1(X, \mathscr{F}) & \to 0 \\ & \downarrow & & \downarrow & & \downarrow & & & \\ 0 \to & \Gamma(X, \mathscr{F}) & \to & \Gamma(X, \mathscr{C}^0) & \to & \Gamma(X, \mathscr{Z}^1) & \to & \tilde{H}^1(X, \mathscr{F}) & \to 0 \end{array}$$

where the 0's on the right come from the fact that $H^1(X, \mathscr{C}^0) = \tilde{H}^1(X, \mathscr{C}^0) = 0$ since both groups satisfy the conditions of theorem 2.8. Because of the isomorphism α_0, the three vertical arrows are identities. It follows that there is a unique isomorphism $\alpha_1: H^1(X, \mathscr{F}) \to \tilde{H}^1(X, \mathscr{F})$ such that the diagram commutes.

By proceeding in the same way and using theorem 2.7, one proves the existence of homomorphisms

$$\alpha_n : H^n(X, \mathscr{F}) \to \tilde{H}^n(X, \mathscr{F})$$

which commute with the homomorphisms δ_n and $\tilde{\delta}_n$. ○

Once the existence and uniqueness of the cohomology groups $H^n(X, \mathscr{F})$ is established, there are several methods of computing them. A simple one is shown in the next theorem.

2.10 THEOREM *Let $\mathscr{F}$ be a sheaf of abelian groups over X. Let*

$$0 \to \mathscr{F} \xrightarrow{\varepsilon} \mathscr{E}^0 \xrightarrow{d_0} \mathscr{E}^1 \xrightarrow{d_1} \mathscr{E}^2 \to \cdots$$

be a soft resolution of $\mathscr{F}$, *and let*

$$0 \to \Gamma(X, \mathscr{F}) \xrightarrow{\varepsilon^*} \Gamma(X, \mathscr{E}^0) \xrightarrow{d_0^*} \Gamma(X, \mathscr{E}^1) \xrightarrow{d_1^*} \Gamma(X, \mathscr{E}^2) \to \cdots$$

be the corresponding sequence of cross-sections. Then we have:

$$H^0(X, \mathscr{F}) = \operatorname{Ker} d_0^*$$

anf for $n \geq 1$

$$H^n(X, \mathscr{F}) = \operatorname{Ker} d_n^* / \operatorname{Im} d_{n-1}^* .$$

Proof That $H^0(X, \mathscr{F}) = \operatorname{Ker} d_0^*$ follows at once from the fact that the sequence

$$0 \to \Gamma(X, \mathscr{F}) \xrightarrow{\varepsilon^*} \Gamma(X, \mathscr{E}^0) \xrightarrow{d_0^*} \Gamma(X, \mathscr{E}^1)$$

is exact.

For $k \geq 0$ the exactness of the resolution of $\mathscr{F}$ is equivalent to the exactness of the sequence of sheaves

$$0 \to \mathscr{K}er\, d_k \to \mathscr{E}^k \xrightarrow{p_k} \mathscr{K}er\, d_{k+1} \to 0 .$$

By theorem 2.6, we then have an exact sequence of cohomology groups

$$\begin{aligned} 0 &\to H^0(X, \mathscr{K}er\, d_k) \to H^0(X, \mathscr{E}^k) \xrightarrow{p_k} H^0(X, \mathscr{K}er\, d_{k+1}) \\ &\to H^1(X, \mathscr{K}er\, d_k) \to H^1(X, \mathscr{E}^k) \to H^1(X, \mathscr{K}er\, d_{k+1}) \\ &\to H^2(X, \mathscr{K}er\, d_k) \to \cdots . \end{aligned}$$

Now, by theorem 2.8, we have $H^n(X, \mathscr{E}^k) = 0$ if $n \geq 1$ since $\mathscr{E}^k$ is a soft sheaf. It follows that for $n \geq 1$ we have isomorphisms

$$H^{n+1}(X, \mathscr{K}er\, d_k) \simeq H^n(X, \mathscr{K}er\, d_{k+1})$$

and

$$H^1(X, \mathscr{K}er\, d_k) \simeq H^0(X, \mathscr{K}er\, d_{k+1}) / p_k H^0(X, \mathscr{E}^k).$$

Then, since $\mathscr{F}$ is isomorphic to $\mathscr{K}er\, d_0$, we have

$$\begin{aligned} H^n(X, \mathscr{F}) &\simeq H^n(X, \mathscr{K}er\, d_0) \simeq H^{n-1}(X, \mathscr{K}er\, d_1) \simeq \cdots \\ \cdots &\simeq H^1(X, \mathscr{K}er\, d_{n-1}) \simeq H^0(X, \mathscr{K}er\, d_n) / p_{n-1} H^0(X, \mathscr{E}^{n-1}) \\ &\simeq \operatorname{Ker} d_n^* / \operatorname{Im} d_{n-1}^* . \quad \bigcirc \end{aligned}$$

The above theorem is known as the abstract de Rahm theorem. In the following sections, we shall show two particular examples of this theorem which are of great importance in the theory of manifolds.

3.1 Poincaré's lemma

Let M be an n-dimensional differentiable manifold and let ω be a differential form defined on M. We know that, if ω is exact, then ω is closed. The converse is not true in general. However, it is true locally, as the next lemma will show; that is, every point $m \in M$ has a neighborhood U such that every closed differential form defined on U is exact in U. In order that this result be globally true, M must satisfy some topological conditions which we shall investigate later.

We consider the following open subset of $\mathbf{R}^n$:

$$D = \{x \in \mathbf{R}^n \mid |x_i| < r_i;\ r_i > 0;\ i = 1, \ldots, n\}.$$

3.2 LEMMA (Poincaré) *Let ω be a closed differential form of degree $p \geq 1$ defined on D. There exists a differential form α of degree $p - 1$ defined on D such that $d\alpha = \omega$.*

Proof We recall first that, if $f(x)$ is a differentiable function of the single variable x defined in the open interval $|x| < r$, $r > 0$, there exists a function

$$g(x) = \int_0^x f(t)\, dt,$$

defined in the same interval, such that

$$\frac{dg\,(x)}{dx} = f(x).$$

Moreover, if f depends differentiably on certain parameters, the same is true for g.

A differential form ω of degree p in D has the expression

3.3 $$\omega = \Sigma\, \omega_{i_1 \cdots i_p} dx^{i_1} \wedge \cdots \wedge dx^{i_p}.$$

Let Q_j be the set of all differential forms of degree p such that the expression 3.3 involves only the differentials $dx^1, \ldots, dx^j$.

We shall show first that the lemma is true for $j = 1$. If $\omega \in Q_1$, since ω has positive degree, it must have the expression

$$\omega = f(x_1, \ldots, x_n)\, dx^1.$$

Since $d\omega = 0$, the partial derivatives $\dfrac{\partial f}{\partial x_2}, \ldots, \dfrac{\partial f}{\partial x_n}$ are all equal to zero; that is, the function f depends only on the variable x_1. It follows that there exists

a function g which also depends only on x_1 such that $f(x_1, \ldots, x_n) = \dfrac{\partial g(x)}{\partial x_1}$.
Hence, by putting $\alpha = g$, we have

$$d\alpha = \frac{\partial g(x)}{\partial x_1} dx^1 = \omega .$$

We now prove the lemma by induction. Assuming the lemma true for Q_{j-1}, let $\omega \in Q_j$. Then ω can be written as

$$\omega = dx^j \wedge \sigma + \tau$$

with $\sigma, \tau \in Q_{j-1}$.

From $d\omega = 0$ it follows that the coefficients of σ and τ are independent of the variables $x_{j+1}, \ldots, x_n$.

Let $f_{i_1 \cdots i_{p-1}}$ be the coefficients of the differential form σ. Denote by $g_{i_1 \cdots i_{p-1}}$ the functions satisfying the relation

$$f_{i_1 \cdots i_{p-1}} = \frac{\partial g_{i_1 \cdots i_{p-1}}}{\partial x_j}$$

and consider the differential form

$$\beta = \Sigma\, g_{i_1 \cdots i_{p-1}} dx^{i_1} \wedge \cdots \wedge dx^{i_{p-1}} .$$

Since the functions $g_{i_1 \cdots i_{p-1}}$ are independent of $x_{j+1}, \ldots, x_n$, we have

$$d\beta = dx^j \wedge \sigma + \tilde{\tau}$$

with $\tilde{\tau} \in Q_{j-1}$. Thus,

$$\omega = d\beta + \tau - \tilde{\tau} .$$

By the induction hypothesis, there exists a differential form γ such that $d\gamma = \tau - \tilde{\tau}$. Hence, $\omega = d\beta + d\gamma$. By putting $\alpha = \beta + \gamma$, we have

$$\omega = d\alpha .$$

This concludes the proof. ○

3.4 *Remark* The differential form α with the property that $d\alpha = \omega$, whose existence is proved in the above lemma, is not uniquely defined.

3.5 *Remark* Every differential form of degree n on D is exact. In fact, every differential form of degree n is closed.

4.1 de Rahm's theorem

Let M be an n-dimensional differentiable manifold and let $\mathscr{A}^p$ be the sheaf of germs of differential forms of degree p defined on M.

If $f \in \mathscr{A}^0$, that is, if f is a differentiable function near every $m \in M$ and $df = 0$, then f is a constant near m. From this and from Poincaré's lemma, it follows that the sequence of sheaves over M

$$0 \to \mathbf{R} \xrightarrow{\varepsilon} \mathscr{A}^0 \xrightarrow{d_0} \mathscr{A}^1 \xrightarrow{d_1} \mathscr{A}^2 \to \cdots \to \mathscr{A}^n \to 0$$

is a resolution of the constant sheaf $\mathbf{R}$ over M. This resolution is soft, since, for every $p \geq 0$, $\mathscr{A}^p$ is a soft sheaf, as it was shown in example 7.12.6. Hence, we can use theorem 2.10 to compute the cohomology groups with real coefficients of the manifold M; that is, we have:

4.2 THEOREM (de Rahm) *Let M be a differentiable manifold. Then for* $p \geq 1$

$$H^p(M, \mathbf{R}) = \frac{\operatorname{Ker}(\Gamma(M, \mathscr{A}^p) \xrightarrow{d_p} \Gamma(M, \mathscr{A}^{p+1}))}{\operatorname{Im}(\Gamma(M, \mathscr{A}^{p-1}) \xrightarrow{d_{p-1}} \Gamma(M, \mathscr{A}^p))}$$

and

$$H^0(M, \mathbf{R}) = \operatorname{Ker}(\Gamma(M, \mathscr{A}^0) \xrightarrow{d_0} \Gamma(M, \mathscr{A}^1)). \quad \bigcirc$$

We recall that the global cross-sections of $\mathscr{A}^p$ over M are differential forms of degree p defined on all M. Then the cohomology group $H^p(M, \mathbf{R})$ is isomorphic to the factor group of the group of closed differential forms of degree p on M by the subgroup of exact differential forms of degree p on M. Hence, on a differentiable manifold M, the two following conditions are equivalent:

a) every closed differential form of degree p on M is exact,

b) the cohomology group $H^p(M, \mathbf{R}) = 0$.

It follows that, given on M any differential form ω with $d\omega = 0$, the existence of a solution to the equation $\omega = d\eta$ depends only on the topological structure of M.

5.1 Dolbeault's lemma

It is clear that Poincaré's lemma still holds if the differential forms we considered in section 3.1 are complex valued. However, the lemma is valid always with respect to the exterior differential d. We shall now show that a similar lemma relative to the differential $\bar{\partial}$ holds for differential forms defined on a subset of $\mathbf{C}^n$.

We shall need the following lemma:

5.2 LEMMA *Let $f(z)$ be a complex valued differentiable function defined on the open disc $D = \{z \in \mathbf{C} \mid |z| < r, r > 0\}$.*

Then for every $r' < r$ there exists a complex valued differentiable function $g(z)$ defined on the disc $D' = \{|z| < r'\}$ such that

$$\frac{\partial g(z)}{\partial \bar{z}} = f(z)$$

on D'.

Proof We recall that, if Ω is a connected open set of $\mathbf{R}^2$, Ω' is a relatively compact open subset of Ω and $f(x, y)$ is a differentiable function on Ω, the differential equation

$$\text{5.3} \qquad \frac{\partial^2 h(x, y)}{\partial x^2} + \frac{\partial^2 h(x, y)}{\partial y^2} = 4 f(x, y)$$

always has a solution in Ω'. The solution h is given by Poisson's integral

$$h(x, y) = \frac{2}{\pi} \int_{\Omega'} f(u, v) \quad \log \varrho \; du \wedge dv,$$

where $\varrho = [(x - u)^2 + (y - v)^2]^{1/2}$.

If now Ω is a subset of $\mathbf{C}$ putting $z = x + iy$ and $w = u + iv$, we have

$$\varrho = |z - w| = [(z - w)(\bar{z} - \bar{w})]^{1/2},$$

$$\frac{\partial h}{\partial x^2} + \frac{\partial h}{\partial y^2} = \frac{1}{4} \frac{\partial^2 h}{\partial z \partial \bar{z}},$$

and

$$du \wedge dv = \frac{i}{2} dw \wedge d\bar{w}.$$

Then the differential equation 5.3 becomes

$$\frac{\partial^2 h(z)}{\partial z \partial \bar{z}} = f(z),$$

and the solution h on Ω' has the expression

$$h(z) = \frac{i}{\pi} \int_{\Omega'} f(w) \log |z - w| \, dw \wedge d\bar{w}.$$

The proof of the lemma now follows by taking $\Omega = D$, $\Omega' = D'$ and by defining

$$g(z) = \frac{\partial h(z)}{\partial z}. \quad \bigcirc$$

If the function $f(z)$ depends differentiably or holomorphically on some parameters, the same is true for the function $g(z)$, since this property holds for Poisson's integral formula.

We now consider the two following polidiscs in $\mathbf{C}^n$:

$$\Delta = \{z \in \mathbf{C}^n \mid |z_j| < r_j; \quad r_j > 0; \quad j = 1, \ldots, n\}$$

and $\Delta' \subset \Delta$,

$$\Delta' = \{z \in \mathbf{C}^n \mid |z_j| < r'_j; \quad 0 < r'_j < r_j; \quad j = 1, \ldots, n\}.$$

5.4 Lemma (Dolbeault) *Let ω be a $\bar{\partial}$-closed differential form of type (p, q) with $q \geq 1$ defined on Δ. There exists a differential form α of type $(p, q - 1)$ defined on Δ' such that*

$$\omega = \bar{\partial}\alpha$$

on Δ'.

Proof The proof is by induction and similar to that of the Poincaré lemma. We denote by Q_j the set of all differential forms of type (p, q) whose coordinate expression does not involve the differentials $d\bar{z}^{j+1}, \ldots, d\bar{z}^n$.

Let $\omega^{p,q} \in Q_1$; since $q \geq 1$, such a form must have the expression

$$\omega^{p,1} = d\bar{z}^1 \wedge \sigma^{p,0}$$

with

$$\sigma^{p,0} = \Sigma f_{i_1 \cdots i_p} \, dz^{i_1} \wedge \cdots \wedge dz^{i_p}.$$

Since $\bar{\partial}\omega^{p,1} = 0$, it follows that all coefficients of $\bar{\partial}\sigma^{p,0}$ are equal to zero, except possibly the coefficient containing the partial derivatives of the coefficients of $\sigma^{p,0}$ with respect to $\bar{z}_1$; that is, the coefficients $f_{i_1 \cdots i_p}$ of $\sigma^{p,0}$ depend holomorphically on the variables $z_2, \ldots, z_n$.

By lemma 5.2, there exist functions $g_{i_1 \cdots i_p}$ such that

$$f_{i_1 \cdots i_p} = \frac{\partial g_{i_1 \cdots i_p}}{\partial \bar{z}_1}.$$

Then the differential form

$$\alpha^{p,0} = \Sigma g_{i_1 \cdots i_p} dz^{i_1} \wedge \cdots \wedge dz^{i_p}$$

has the property that $\bar{\partial}\alpha^{p,0} = \omega^{p,1}$, and this proves the lemma for $j = 1$. We now assume the lemma true for Q_{j-1} and let $\omega^{p,q} \in Q_j$. Then we can write $\omega^{p,q}$ as

$$\omega^{p,q} = d\bar{z}^j \wedge \sigma^{p,q-1} + \tau^{p,q}$$

with $\sigma^{p,q-1}, \tau^{p,q} \in Q_{j-1}$.

From $\bar{\partial}\omega^{p,q} = 0$ it follows that the coefficients $f_{i_1 \cdots i_p j_1 \cdots j_{q-1}}$ of $\sigma^{p,q-1}$ are holomorphic functions of the variables $z_{j+1}, \ldots, z_n$. Again by lemma 5.2, there exist functions $g_{i_1 \cdots i_p j_1 \cdots j_{q-1}}$ such that

$$f_{i_1 \cdots i_p j_1 \cdots j_{q-1}} = \frac{\partial g_{i_1 \cdots i_p j_1 \cdots j_{q-1}}}{\partial \bar{z}_j}.$$

We consider the differential form

$$\beta^{p,q-1} = \Sigma g_{i_1 \cdots i_p j_1 \cdots j_{q-1}} dz^{i_1} \wedge \cdots \wedge dz^{i_p} \wedge d\bar{z}^{j_1} \wedge \cdots \wedge d\bar{z}^{j_{q-1}}$$

and compute $\bar{\partial}\beta^{p,q-1}$. Since the coefficients $g_{i_1 \cdots i_p j_1 \cdots j_{q-1}}$ are holomorphic functions of the variables $z_{j+1}, \ldots, z_n$, we have

$$\bar{\partial}\beta^{p,q-1} = d\bar{z}^j \wedge \sigma^{p,q-1} + \tilde{\tau}^{p,q}$$

with $\tilde{\tau}^{p,q} \in Q_{j-1}$. Thus,

$$\omega^{p,q} = \bar{\partial}\beta^{p,q-1} + \tau^{p,q} - \tilde{\tau}^{p,q}$$

with $\tau^{p,q} - \tilde{\tau}^{p,q} \in Q_{j-1}$. By assumption there exists a differential form $\gamma^{p,q-1}$ such that $\bar{\partial}\gamma^{p,q-1} = \tau^{p,q} - \tilde{\tau}^{p,q}$. Hence,

$$\omega^{p,q} = \bar{\partial}\beta^{p,q-1} + \bar{\partial}\gamma^{p,q-1}.$$

By putting $\alpha^{p,q-1} = \beta^{p,q-1} + \gamma^{p,q-1}$, we have

$$\omega^{p,q} = \bar{\partial}\alpha^{p,q-1},$$

which proves the lemma. ○

5.5 *Remark* As in the Poincaré lemma, the differential form $\alpha^{p,q-1}$, such that $\omega^{p,q} = \bar{\partial}\alpha^{p,q-1}$, whose existence is proved in the Dolbeault lemma, is not uniquely defined.

5.6 *Remark* Let $\omega^{p,n}$ be any differential form defined on a neighborhood of the closure of a polidisc Δ in $\mathbf{C}^n$. Then $\omega^{p,n}$ is $\bar{\partial}$-exact on Δ. In fact, every differential form of type (p, n) is $\bar{\partial}$-closed.

6.1 Dolbeault's theorem

Let M be an n-dimensional complex manifold and let $\mathscr{E}^{p,q}$ be the sheaf of germs of differential forms of type (p, q) on M. For every $0 \leq p, q \leq n$, the sheaf $\mathscr{E}^{p,q}$ is soft by example 7.12.6. Let $\mathcal{O}$ be the sheaf of germs of holomorphic functions on M. Then the sequence of sheaves over M

$$0 \to \mathcal{O} \xrightarrow{\varepsilon} \mathscr{E}^{0,0} \xrightarrow{\bar{\partial}_0} \mathscr{E}^{0,1} \xrightarrow{\bar{\partial}_1} \mathscr{E}^{0,2} \xrightarrow{\bar{\partial}_2} \cdots \to \mathscr{E}^{0,n} \to 0$$

is a soft resolution of the sheaf $\mathcal{O}$. To prove the exactness at $\mathscr{E}^{0,0}$, we remark that, if $\mathscr{f} \in \mathscr{E}^{0,0}$ at any point $m \in M$ and $\bar{\partial}\mathscr{f} = 0$, then there exists a representative f of the germ $\mathscr{f}$ which is a holomorphic function on some neighborhood of m. Thus, $\mathcal{O} = \mathscr{K}er\, \bar{\partial}_0$.

The exactness at $\mathscr{E}^{0,p}$ for $p \geq 1$ follows from Dolbeault's lemma.

Hence, we can use theorem 2.10 to compute the cohomology groups with coefficients in the sheaf $\mathcal{O}$ of the complex manifold M; that is, we have

6.2 THEOREM (Dolbeault) *Let M be a complex manifold. Then, for p $\geq$ 1,*

$$H^p(M, \mathcal{O}) = \frac{\operatorname{Ker}(\Gamma(M, \mathscr{E}^{0,p}) \xrightarrow{\bar{\partial}_p} \Gamma(M, \mathscr{E}^{0,p+1}))}{\operatorname{Im}(\Gamma(M, \mathscr{E}^{0,p-1}) \xrightarrow{\bar{\partial}_{p-1}} \Gamma(M, \mathscr{E}^{0,p}))}$$

and

$$H^0(M, \mathcal{O}) = \operatorname{Ker}(\Gamma(M, \mathscr{E}^{0,0}) \xrightarrow{\bar{\partial}_0} \Gamma(M, \mathscr{E}^{0,1})). \bigcirc$$

More generally, let us denote by $\Omega^{p,0}$ the sheaf of germs of holomorphic differential forms of type $(p, 0)$ on M. Then we have an exact sequence of sheaves over M

$$0 \to \Omega^{p,0} \xrightarrow{\varepsilon} \mathscr{E}^{p,0} \xrightarrow{\bar{\partial}} \mathscr{E}^{p,1} \xrightarrow{\bar{\partial}} \mathscr{E}^{p,2} \xrightarrow{\bar{\partial}} \cdots \to \mathscr{E}^{p,n} \to 0,$$

which is a soft resolution of the sheaf $\Omega^{p,0}$. The same argument as before now yields

6.3
$$H^q(M, \Omega^{p,0}) = \frac{\operatorname{Ker}(\Gamma(M, \mathscr{E}^{p,q}) \xrightarrow{\bar{\partial}} \Gamma(M, \mathscr{E}^{p,q+1}))}{\operatorname{Im}(\Gamma(M, \mathscr{E}^{p,q-1}) \xrightarrow{\bar{\partial}} \Gamma(M, \mathscr{E}^{p,q}))}.$$

The groups appearing on the right hand side of 6.3 are usually called *Dolbeault cohomology groups* and are denoted by $H^{p,q}(M)$. They consist of $\bar{\partial}$-closed differential forms of type (p, q) on M modulo $\bar{\partial}$-exact differential forms of type (p, q) on M. Using this notation, we can reformulate Dolbeault's theorem in its more general form as follows:

6.4 THEOREM *Let M be a complex manifold; then*

$$H^q(M, \Omega^{p,0}) \simeq H^{p,q}(M). \ \bigcirc$$

In particular, since $\Omega^{0,0} = \mathcal{O}$, we have $H^q(M, \mathcal{O}) \simeq H^{0,q}(M)$.

7.1 Problems

1 (1.1) Complete the proof of theorem 1.6.

2 (2.1) Fill in all missing details in the proof of theorem 2.9.

CHAPTER 9

Kähler manifolds

1.1 Adjoint form

LET M BE an n-dimensional orientable differentiable manifold. We shall assume that M is oriented and that a Riemannian metric is defined on M. Riemannian metrics were considered in 6.7.2, and it was proved that the assignment of a Riemannian metric on M is equivalent to a reduction of the structure group of the cotangent bundle $T^*(M)$ to the orthogonal group.

1.2 DEFINITION *A differentiable manifold M is called a Riemann manifold if a Riemannian metric is defined on M.*

In a local chart (U_α, u_α) of a Riemann manifold M, the metric has the expression

$$ds_\alpha^2 = \sum_{i=1}^{n} (\omega_\alpha^i)^2$$

where the differential forms $\omega_\alpha^1, \ldots, \omega_\alpha^n$ are ordered so that the differential form $\omega_\alpha^1 \wedge \cdots \wedge \omega_\alpha^n$ is positive with respect to the fixed orientation of M.

The differential forms ω_α^i are said to be a *system of structure forms* of M on U_α.

We remark that, in general, the structure forms ω_α^i are not the differentials of the coordinates of any system of coordinates on U_α. If in every local chart the ω_α^i can be chosen to be the differentials of local coordinates, the manifold M is called *locally Euclidean*.

Let $A^p(U_\alpha) = \Gamma_d(U_\alpha, \mathscr{A}^p(M))$. The Riemannian metric on M allows us to define an operator

$$*: A^p(U_\alpha) \to A^{n-p}(U_\alpha)$$

from differential forms of degree p on U_α to differential forms of degree $n-p$ on U_α, in the following way: to the differential form $\omega_\alpha^{i_1} \wedge \cdots \wedge \omega_\alpha^{i_p}$ we associate the differential form of degree $n-p$,

$$* (\omega_\alpha^{i_1} \wedge \cdots \wedge \omega_\alpha^{i_p}) = \operatorname{sign}(i_1 \cdots i_p j_1 \cdots j_{n-p})\, \omega_\alpha^{j_1} \wedge \cdots \wedge \omega_\alpha^{j_{n-p}},$$

where $i_1 \cdots i_p j_1 \cdots j_{n-p}$ is a permutation of $1 \cdots n$.

We then extend the operator $*$, by linearity, to every differential form of degree p on U_α. Here, by linearity we mean linearity with respect to differentiable functions on U_α.

The operator $*$ is defined locally, but it is actually independent of the local coordinates.

1.3 THEOREM *The operator*

$$*: A^p(M) \to A^{n-p}(M)$$

depends only on the Riemannian metric and on the orientation of M.

Proof It follows at once from the fact that on $U_\alpha \cap U_\beta$ the ω_α^i are related to the ω_β^i by an orthogonal transformation. ○

It follows from the definition that the operator $*$ is an isomorphism of the vector space $A^p(M)$ with the vector space $A^{n-p}(M)$.

If $\eta^p \in A^p(M)$, the differential form $*\eta^p$ is called the *adjoint* of η^p.

An easy computation shows that

$$**\eta^p = (-1)^{p(n-p)}\eta^p.$$

Let the Riemannian metric on M be given by

$$ds_\alpha^2 = \sum_{i,j=1}^{n} g_{ij} du_\alpha^i du_\alpha^j$$

on the coordinate neighborhood U_α. At every point $m \in U_\alpha$ the $g_{ij}(m)$ are the components of a covariant tensor of order two. We shall denote by $g^{ij}(m)$ the components of the contravariant tensor of order two defined by the equations

$$\sum_{k=1}^{n} g^{ik}(m)\, g_{kj}(m) = \delta_j^i.$$

If a differential form $\omega^p \in A^p(U_\alpha)$ is given by

$$\omega^p = \sum_{1 \le i_1 < \cdots < i_p \le n} \omega_{i_1 \cdots i_p} du_\alpha^{i_1} \wedge \cdots \wedge du_\alpha^{i_p},$$

the coefficients $\omega_{i_1 \cdots i_p}$ of ω^p are the components of a covariant tensor of order p at $m \in U_\alpha$. They define a contravariant tensor of order p by

$$\omega^{i_1 \cdots i_p} = \sum_{j_1 \cdots j_p} g^{i_1 j_1} \cdots g^{i_p j_p} \omega_{j_1 \cdots j_p}.$$

We put $g = \det(g_{ij})$. The differential form of degree n,

$$\sqrt{g}\, du^1 \wedge \cdots \wedge du^n,$$

where $\sqrt{g} > 0$, is called the *volume element* of M. It is clearly independent of the local coordinates.

In these conditions, the adjoint form of ω^p has the expression

$$*\omega^p = \sum_{1 \leq j_1 < \cdots < j_{n-p} \leq n} \operatorname{sign}(i_1 \cdots i_p j_1 \cdots j_{n-p}) \sqrt{g}\, \omega^{i_1 \cdots i_p} du_\alpha^{j_1} \wedge \cdots \wedge du_\alpha^{j_{n-p}}.$$

It is easy to check that the operator $*$ is defined in such a way as to satisfy the following conditions:

i) the subspace of $*\omega^p$ is orthogonal to the subspace of ω^p,

ii) the $(n - p)$-dimensional volume of $*\omega^p$ is equal to the p-dimensional volume of ω^p,

iii) the differential form $\omega^p \wedge *\omega^p$ is positive with respect to the orientation of M,

iv) the adjoint of the differential form of degree zero, equal to the constant 1, is the volume element; that is,

$$*1 = \sqrt{g}\, du^1 \wedge \cdots \wedge du^n,$$

v) $*$ is a linear operator; that is, for any differentiable functions f, g and differential forms ω, η,

$$*(f\omega + g\eta) = f * \omega + g * \eta,$$

vi) for every pair of differential forms ω^p, η^p, of the same degree p,

$$\omega^p \wedge *\eta^p = \eta^p \wedge *\omega^p.$$

2.1 Inner product

Let M be a Riemann manifold of dimension n. We shall denote by $D^p(M)$ the subspace of $A^p(M)$ of differential forms of degree p with compact support in M.

If ω^p, η^p are in $A^p(M)$, we can consider the differential form $\omega^p \wedge *\eta^p \in A^n(M)$ and the integral

$$\int_M \omega^p \wedge *\eta^p.$$

If this integral exists and it is finite, it is called the *inner product* of the differential forms ω^p, η^p and is denoted by

$$(\omega^p, \eta^p) = \int_M \omega^p \wedge *\eta^p.$$

The inner product of two differential forms of the same degree may not be defined. If it is defined (ω^p, η^p) satisfies the following properties:

2.2
$$\begin{cases} \text{i) } (\omega^p, \eta^p) = (\eta^p, \omega^p), \\ \text{ii) } (\omega^p, \eta^p) \text{ is bilinear,} \\ \text{iii) } (\omega^p, \omega^p) \geq 0 \text{ for every } \omega^p \in A^p(M), \\ \text{iv) } (\omega^p, \omega^p) = 0 \text{ if and only if } \omega^p = 0. \end{cases}$$

These properties follow at once from the definition of the operator $*$.

If ω, η are nonhomogeneous differential forms, that is, if $\omega, \eta \in A(M) = \bigoplus_{p=0}^{n} A^p(M)$, the definition of the inner product extends to $A(M)$ by putting

$$(\omega, \eta) = \sum_{p=0}^{n} (\omega^p, \eta^p),$$

where $\omega^p, \eta^p \in A^p(M)$. Properties 2.2 clearly generalize to

j) $(\omega, \eta) = (\eta, \omega)$,
jj) (ω, η) is bilinear,
jjj) $(\omega, \omega) \geq 0$,
jv) $(\omega, \omega) = 0$ if and only if $\omega = 0$.

If $\omega, \eta \in D(M) = \bigoplus_{p=0}^{n} D^p(M)$, the inner product (ω, η) is always defined. Thus, the space $D(M)$ of differential forms with compact support in M is a pre-Hilbert space.

We remark that, if the manifold M is compact, then $A(M) = D(M)$; that is, every differential form has compact support.

We shall denote by $L^2_{(p)}(M)$ the space of all differential forms of degree p, such that the coefficients are measurable in any local coordinate system and square integrable with respect to the metric of M. If we denote by $\overline{D^p(M)}$ the Hilbert space obtained by completion of $D^p(M)$, then we have

$$L^2_{(p)}(M) = A^p(M) \cap \overline{D^p(M)}.$$

If $\omega^p \in L^2_{(p)}(M)$, the norm of ω^p is defined by

$$||\omega^p|| = (\omega^p, \omega^p)^{1/2}.$$

Let $\omega, \eta \in A(M)$. The differential form η is said to be *orthogonal* to ω if (ω, η) is defined and $(\omega, \eta) = 0$.

2.3 LEMMA *If $\eta \in A(M)$ is orthogonal to every $\omega \in D(M)$, then $\eta = 0$.*

Proof Let $m \in M$ and assume that $\eta(m) \neq 0$. Let K be a compact set containing m, let U be an open neighborhood of K in M, and let f be a differentiable function on M such that

$$f(m') \geq 0 \quad \text{on} \quad M,$$

$$f(m') = 1 \quad \text{if} \quad m' \in K,$$

$$f(m') = 0 \quad \text{if} \quad m' \in M \backslash U.$$

The differential form $f\eta \in D(M)$ and

$$(\eta, f\eta) > 0.$$

This contradicts the hypothesis that $(\eta, \omega) = 0$ for every $\omega \in D(M)$. ○

3.1 Adjoint operators

Let $T : A(M) \to A(M)$ be a linear operator.

3.2 DEFINITION *A linear operator $T' : D(M) \to A(M)$ is called an adjoint of T if, for every pair of differential forms $\omega, \eta \in D(M)$,*

$$(T\omega, \eta) = (\omega, T'\eta).$$

The adjoint T' of a linear operator T is sometimes called a metric transpose of T.

3.3 THEOREM *Let $T: A(M) \to A(M)$ be a linear operator. If an adjoint T' of T exists, it is unique.*

Proof Let $T', \tilde{T}'$ be two adjoints of T. We shall prove that, for every $\eta \in D(M)$, $(T' - \tilde{T}')\eta = 0$. Let ω, η be any two elements of $D(M)$. By hypothesis,

$$(T\omega, \eta) = (\omega, T'\eta) = (\omega, \tilde{T}'\eta).$$

Thus,

$$(\omega, (T' - \tilde{T}')\eta) = 0.$$

The differential form $(T' - \tilde{T}')\eta$ is an element of $A(M)$, orthogonal to every $\omega \in D(M)$, and, thus, by lemma 2.3, $(T' - \tilde{T}')\eta = 0$. ○

3.4 DEFINITION *A linear operator $T: A(M) \to A(M)$ is said to be a local operator if, for every differential form $\omega \in A(M)$,*

$$\text{Supp}\, T\omega \subset \text{Supp}\, \omega.$$

Let ω, $\tilde{\omega}$ be two differential forms representing the same germ of $\mathscr{A}_m$ at some point $m \in M$. If T is a local operator, $T\omega$ and $T\tilde{\omega}$ clearly determine the same germ. In fact, if this were not true, we would have that $m \notin \text{Supp}\,(\omega - \tilde{\omega})$, while, $m \in \text{Supp}\, T(\omega - \tilde{\omega})$, contrary to the hypothesis that T is a local operator. Conversely, every linear operator defined on germs of differential forms is clearly local. It follows that the notion of a local operator is exactly the same as that of a linear operator on germs of differential forms.

3.5 THEOREM *Let $T: A(M) \to A(M)$ be a local operator. If the adjoint $T': D(M) \to A(M)$ exists, then T' is also local.*

Proof If T' is not a local operator, there exist a differential form $\omega \in D(M)$ and a point $m \in M$ such that $m \in \text{Supp}\, T'\omega$ and $m \notin \text{Supp}\, \omega$. Then we can find a differentiable function f such that $f \geq 0$ on M, $f(m) = 1$, and $f = 0$ outside a compact set disjoint from $\text{Supp}\, \omega$.

Consider the differential form $\eta = fT'\omega \in D(M)$. We have

$$(T'\omega, \eta) = (T'\omega, fT'\omega) > 0.$$

We now observe that $\text{Supp}\, \omega \cap \text{Supp}\, \eta = \emptyset$. Moreover, since T is local, $\text{Supp}\, T\eta \subset \text{Supp}\, \eta$. It follows that

$$(T'\omega, \eta) = (\omega, T\eta) = 0.$$

This contradiction proves the theorem. ○

3.6 *Remark* The adjoint T' of a linear operator T is defined on $D(M)$. Sometimes T' may be extended to an operator $\hat{T}': A(M) \to A(M)$; that is, $\hat{T}'$ restricted to $D(M)$ coincides with T'. For instance, if T is a local operator and T' exists, there is a unique operator $\hat{T}'$ which is equal to T' on $D(M)$. In fact, let $\omega \in A(M)$. To every point $m \in M$ we can find a differential form $\eta \in D(M)$ such that ω and η belong to the same germ at m. We then define $\hat{T}'\omega\,(m) = T'\eta\,(m)$.

3.7 *Remark* Given two differential forms $\omega, \eta \in A(M)$, the inner products $(T\omega, \eta)$ and $(\omega, \hat{T}'\eta)$ may both exist even if neither ω or η are elements of $D(M)$. However, if we do not assume that $\omega, \eta \in D(M)$, then it does not necessarily follow that $(T\omega, \eta) = (\omega, \hat{T}'\eta)$. It is easy to prove that, if T is local, the equality holds if either ω or η are elements of $D(M)$.

3.8 *Examples* The linear operators

$$d: A^p(M) \to A^{p+1}(M),$$

and

$$*: A^p(M) \to A^{n-p}(M),$$

which we have already considered, are examples of local operators. The operator $*$ is self-adjoint up to the sign; that is, the adjoint of $*$ is $*$ itself.

The adjoint of $d: A^{p-1}(M) \to A^p(M)$ is denoted by

$$\delta: A^p(M) \to A^{p-1}(M),$$

and, for $\eta^p \in A^p(M)$, is defined by

$$\delta\eta^p = -(-1)^{n(p+1)} *d*\eta^p.$$

To show that δ is the adjoint of d, we must prove that, for every pair of differential forms $\omega^{p-1}, \eta^p \in D(M)$, we have $(d\omega^{p-1}, \eta^p) = (\omega^{p-1}, \delta\eta^p)$. To do so, we apply Stokes' formula to a domain $\Omega \supset \operatorname{Supp} \eta^p$. We have

$$(d\omega^{p-1}, \eta^p) - (\omega^{p-1}, \delta\eta^p) = \int_\Omega d\omega^{p-1} \wedge *\eta^p + (-1)^{n(p+1)} \int_\Omega \omega^{p-1} \wedge *(*d*\eta^p)$$

$$= \int_\Omega (d\omega^{p-1} \wedge *\eta^p + (-1)^{p-1}\omega^{p-1} \wedge d*\eta^p) = \int_\Omega d(\omega^{p-1} \wedge *\eta^p)$$

$$= \int_{\partial\Omega} \omega^{p-1} \wedge *\eta^p = 0.$$

A differential form ω such that $\delta\omega = 0$ is called *δ-closed*. If $\omega = \delta\eta$ for some η, then ω is called *δ-exact*.

3.9 *Remark* According to remark 3.7, the inner products $(d\omega, \eta)$ and $(\omega, \delta\eta)$ may both exist even if none of the differential forms ω, η are in $D(M)$. However, in this case, if $\{\Omega_i\}$ is a sequence of domains such that $\Omega_i \to M$, then it does not necessarily follow that $\lim_{\Omega_i \to M} \int_{\partial\Omega_i} \omega \wedge *\eta = 0$.

4.1 Laplace operator

Let M be an n-dimensional Riemann manifold. For every p, $0 \leq p \leq n$, we define an operator

$$\Delta : A^p(M) \to A^p(M)$$

by

$$\Delta = d\delta + \delta d.$$

It is easy to show that Δ is a self-adjoint operator. In fact, if $\omega \in A^p(M)$ and $\eta \in D^p(M)$, we have

$$(\Delta\omega, \eta) = (d\delta\omega, \eta) + (\delta d\omega, \eta) = (\omega, d\delta\eta) + (\omega, \delta d\eta) = (\omega, \Delta\eta).$$

It is also easy to check that Δ commutes with the operators $*$, d, δ; that is,

$$*\Delta = \Delta *, \quad d\Delta = \Delta d, \quad \delta\Delta = \Delta\delta.$$

4.2 THEOREM *Let $M = \mathbf{R}^n$ with the natural Euclidean metric. For $p = 0$, the operator Δ coincides up the sign with the Laplace operator on functions.*

Proof If $p = 0$, we first have $\Delta = \delta d = -*d*d$. Next, since

$$df = \sum_{i=1}^{n} \frac{\partial f}{\partial x_i} dx^i,$$

and

$$*dx^i = (-1)^{i-1} dx^1 \wedge \cdots \wedge \widehat{dx^i} \wedge \cdots \wedge dx^n,$$

we have

$$d*df = \sum_{i=1}^{n} \frac{\partial^2 f}{\partial x_i^2} dx^1 \wedge \cdots \wedge dx^n.$$

Therefore,

$$\Delta f = -*d*df = -\sum_{i=1}^{n} \frac{\partial^2 f}{\partial x_i^2}. \quad \bigcirc$$

Because of the previous theorem, the operator Δ is called a generalized Laplace operator and a differential form ω such that $\Delta\omega = 0$ is called *harmonic*.

The set of all harmonic differential forms of degree p on M is a vector space denoted by $\mathbf{H}^p(M)$. We shall also consider the vector space $\mathbf{H}^p_c(M)$ of harmonic forms of degree p whose coefficients are measurable in any local coordinate system and square integrable with respect to the metric of M.

One has

$$\mathbf{H}_c^p(M) = \mathbf{H}^p(M) \cap \overline{D^p(M)}.$$

If the manifold M is compact, then $\mathbf{H}_c^p(M) = \mathbf{H}^p(M)$.

It is clear that, if $\omega \in A(M)$ is any d-closed and δ-closed differential form, then ω is harmonic; that is, the conditions $d\omega = 0$ and $\delta\omega = 0$ imply $\Delta\omega = 0$. The converse is not true in general. In fact, if $\omega \in \mathbf{H}^p(M)$ but $\omega \notin \mathbf{H}_c^p(M)$, then $\Delta\omega = 0$ does not necessarily imply that $d\omega = 0$ and $\delta\omega = 0$. For example, let $M = \mathbf{R}$ with the natural Euclidean metric. Consider the differential form of degree zero, $\omega = x$. We have $\Delta\omega = 0$ but $d\omega \neq 0$.

4.3 THEOREM *Let M be a compact manifold. Then,*

$$\mathbf{H}^p(M) = \{\omega \in A^p(M) \mid d\omega = \delta\omega = 0\}.$$

Proof Let $\omega \in A^p(M)$ and let $\Delta\omega = 0$. Then, $(\Delta\omega, \omega) = 0$. We have

$$0 = (\Delta\omega, \omega) = (d\delta\omega, \omega) + (\delta d\omega, \omega) = (\delta\omega, \delta\omega) + (d\omega, d\omega),$$

which implies

$$(\delta\omega, \delta\omega) = 0 \quad \text{and} \quad (d\omega, d\omega) = 0.$$

Thus,

$$\delta\omega = 0 \quad \text{and} \quad d\omega = 0. \ \bigcirc$$

If M is not compact, the above theorem is still true under the hypothesis that the metric on M is complete. We do not prove the theorem in this more general form. However, we remark that the argument of the previous proof cannot be applied since, if M is not compact, according to remark 3.7, we do not necessarily have $(d\delta\omega, \omega) = (\delta\omega, \delta\omega)$ and $(\delta d\omega, \omega) = (d\omega, d\omega)$.

5.1 Hermitian metric

Let V, W be two finite dimensional complex vector spaces. A map $T: V \to W$ is called *antilinear* if, for $u, v \in V$ and $c \in \mathbf{C}$, one has

$$T(u + v) = T(u) + T(v),$$

$$T(cv) = \bar{c}T(v).$$

A map

$$h: V \times V \to \mathbf{C}$$

is said to be *sesquilinear* if, for every $v \in V$, the map $u \to h(u, v)$ is a linear map $V \to \mathbf{C}$, and, for every $u \in V$, the map $v \to h(u, v)$ is an antilinear map

$V \to \mathbf{C}$. The map h is called *Hermitian* if, for every $u, v \in V$, one has

5.2 $$h(v, u) = \overline{h(u, v)}.$$

A complex valued function defined on $V \times V$, which is sesquilinear and Hermitian, is called a *Hermitian form* on V. If $\dim V = n$, a Hermitian form H on V is represented by an $n \times n$ matrix (h_{jk}) and condition 5.2 is equivalent to

$$h_{jk} = \overline{h_{kj}}, \quad j, k = 1, \ldots, n.$$

Let M be a complex manifold, $\dim_{\mathbf{C}} M = n$. Let $\{U_\alpha\}$ be an open covering of M by coordinate neighborhoods. Let $w_{\alpha 1}, \ldots, w_{\alpha n}$ be local coordinates in U_α.

5.3 DEFINITION *A Hermitian metric on M is the assignment on every local chart (U_α, w_α) of a Hermitian form on $T^*(U_\alpha)$*

$$ds_\alpha^2 = \sum_{j,k=1}^{n} h_{jk}\, dw_\alpha^j\, d\bar{w}_\alpha^k$$

such that:

i) *for every j, k, h_{jk} is a complex valued differentiable function on U_α,*

ii) *the matrix $(h_{jk})_m$ is positive definite at every $m \in U_\alpha$,*

iii) *for every α, β, $ds_\alpha^2 = ds_\beta^2$.*

Condition (i) says that the entries of the matrix (h_{jk}) are differentiable functions of the local coordinates, condition (ii) that $ds_\alpha^2 >$, and condition (iii) that in $U_\alpha \cap U_\beta$ we must have

$$h_{jk}(w_\beta, \bar{w}_\beta) = \sum_{r,s=1}^{n} h_{rs}(w_\alpha, \bar{w}_\alpha) \frac{\partial w_{\alpha r}}{\partial w_{\beta j}} \frac{\partial \bar{w}_{\alpha s}}{\partial \bar{w}_{\beta k}}.$$

The next theorem will show that the assignment of a Hermitian metric on M induces on M the structure of a Riemann manifold.

5.4 THEOREM *Let M be an n-dimensional complex manifold. Every Hermitian metric on M induces a Riemannian metric on M.*

Proof Let $\sum_{j,k=1}^{n} h_{jk}\, dw^j\, d\bar{w}^k$ be the local expression of a given Hermitian metric on M in terms of local coordinates $w_j = u_j + iv_j$, $j = 1, \ldots, n$.

We put

$$(dw, d\bar{w}) = (dw^1, \ldots, dw^n, d\bar{w}^1, \ldots, d\bar{w}^n)$$

and

$$\begin{pmatrix} 0 & h_{jk} \\ \overline{h_{jk}} & 0 \end{pmatrix} = \begin{pmatrix} 0 & \cdots & 0 & h_{11} & \cdots & h_{1n} \\ \vdots & & \vdots & \vdots & & \vdots \\ 0 & \cdots & 0 & h_{n1} & \cdots & h_{nn} \\ \overline{h_{11}} & \cdots & \overline{h_{1n}} & 0 & \cdots & 0 \\ \vdots & & \vdots & \vdots & & \vdots \\ \overline{h_{n1}} & \cdots & \overline{h_{nn}} & 0 & \cdots & 0 \end{pmatrix}.$$

Then, with matrix notation, we have

$$\sum_{j,k=1}^{n} h_{jk}\, dw^j\, d\bar{w}^k = \frac{1}{2}(dw, d\bar{w}) \begin{pmatrix} 0 & h_{jk} \\ \overline{h_{jk}} & 0 \end{pmatrix} \begin{pmatrix} dw \\ d\bar{w} \end{pmatrix}.$$

On the other hand, for $j = 1, \ldots, n$,

$$\begin{pmatrix} dw^j \\ d\bar{w}^j \end{pmatrix} = \begin{pmatrix} 1 & i \\ 1 & -i \end{pmatrix} \begin{pmatrix} du^j \\ dv^j \end{pmatrix}.$$

Thus, with self-explanatory notation, we have

$$\begin{aligned} \sum_{j,k=1}^{n} h_{jk}\, dw^j\, d\bar{w}^k &= \frac{1}{2}(du, dv) \begin{pmatrix} 1 & i \\ 1 & -i \end{pmatrix} \begin{pmatrix} 0 & h_{jk} \\ \overline{h_{jk}} & 0 \end{pmatrix} \begin{pmatrix} 1 & i \\ 1 & -i \end{pmatrix} \begin{pmatrix} du \\ dv \end{pmatrix} \\ &= \frac{1}{2}(du, dv) \begin{pmatrix} h_{jk} + \overline{h_{jk}} & i\,(\overline{h_{jk}} - h_{jk}) \\ -i\,(\overline{h_{jk}} - h_{jk}) & h_{jk} + \overline{h_{jk}} \end{pmatrix} \begin{pmatrix} du \\ dv \end{pmatrix}. \end{aligned}$$

The last expression represents a real symmetric and positive definite quadratic form, and, thus, it defines a Riemannian metric on M, considered as a real analytic manifold. ○

If M is a complex manifold and a Hermitian metric is defined on M, then M is called a *Hermitian manifold.* Because a Hermitian manifold is also a Riemann manifold, the operators $*, \delta, \Delta$ are well defined on Hermitian manifolds. These operators are defined on real differential forms and are extended by linearity to complex differential forms. It is easy to check that, on a complex manifold M, the operator $*$ is bihomogeneous; that is, $*$ maps $E^{p,q}(M)$ into $E^{n-q,n-p}(M)$ if $\dim_{\mathbf{C}} M = n$. Because the topological dimension of M is even, the expression of the adjoint of the operator d, with respect to the Riemannian metric, is

$$\delta = -*d*.$$

Every complex manifold can be given a Hermitian metric. Therefore, if M is a complex manifold, we can also consider the adjoints of the operators ∂, $\bar{\partial}$ with respect to a Hermitian metric. The adjoint of ∂: $E^{p,q}(M) \to E^{p+1,q}(M)$, denoted by

$$\bar{\theta} : E^{p+1,q}(M) \to E^{p,q}(M),$$

is defined by

$$\bar{\theta} = -*\bar{\partial}*.$$

Similarly, the adjoint of the operator $\bar{\partial}$: $E^{p,q}(M) \to E^{p,q+1}(M)$, denoted by

$$\theta : E^{p,q+1}(M) \to E^{p,q}(M),$$

is defined by

$$\theta = -*\partial*.$$

The operators $\bar{\theta}$, θ are now used to define two complex Laplace operators. Namely, we define $\square'$, $\square''$: $E^{p,q}(M) \to E^{p,q}(M)$ by

$$\square' = \partial\bar{\theta} + \bar{\theta}\partial$$

and

$$\square'' = \bar{\partial}\theta + \theta\bar{\partial}.$$

It is easy to check that these operators are self-adjoint and type-preserving.

It is clear that all operators we have defined in this section are local operators.

A differential form $\omega^{p,q}$ is called $\square'$-*harmonic* if $\square'\omega^{p,q} = 0$ and $\square''$-*harmonic* if $\square''\omega^{p,q} = 0$.

6.1 Almost complex structures

Roughly speaking, an almost complex manifold is a differentiable manifold together with a complex structure on the tangent bundle.

6.2 DEFINITION *An almost complex structure on a differentiable manifold M is a vector bundle homomorphism*

$$J : T(M) \to T(M)$$

such that:

a) *for every $m \in M$, $J(T_m) = T_m$,*

b) *on every T_m, $J^2 = -$identity.*

A differentiable manifold together with an almost complex structure will be called an *almost complex manifold.* Almost complex manifolds are characterized by the next theorem, the proof of which we leave to the reader as an exercise.

6.3 THEOREM *A differentiable manifold M is an almost complex manifold if and only if the following conditions are satisfied:*

i) *the topological dimension of M is even; that is,* $\dim_{\mathbf{R}} M = 2n$,

ii) *for every $m \in M$, the tangent space T_m has the structure of an n-dimensional complex vector space,*

iii) *to every point $m \in M$, there are a neighborhood U of m in M and n complex valued C^∞ differential forms $\omega^1, \ldots, \omega^n$ on U such that, for every $p \in U$, the map*

$$t_p \to (\langle \omega^1(p), t_p \rangle, \ldots, \langle \omega^n(p), t_p \rangle)$$

is an isomorphism of T_p onto $\mathbf{C}^n$. ○

The differential forms $\omega^1, \ldots, \omega^n$ are a system of structure forms on a neighborhood of m in M, and they are linearly independent. The structure forms $\{\omega^j\}$ together with the conjugates $\{\bar{\omega}^j\}$ form a basis of the cotangent space T_m^*. Thus, on a neighborhood of $m \in M$, every differential form of degree one on M is a linear combination, with complex valued C^∞ coefficients, of the forms $\omega^j, \bar{\omega}^j; j = 1, \ldots, n$. As in the case of complex manifolds, the transpose of the vector bundle homomorphism $J, J^*\colon T^*(M) \to T^*(M)$ induces a bigradation of the differential forms; that is, if M is an almost complex manifold, for every h, $0 \leq h \leq 2n$, we have

$$E^h(M) = \bigoplus_{p+q=h} E^{p,q}(M).$$

An element $\omega \in \Gamma(E^{p,q})$ is called bihomogeneous of type (p, q), and it has a local expression, in terms of the structure forms, as

$$\omega = \sum_{\substack{1 \leq i_1 < \cdots < i_p \leq n \\ 1 \leq j_1 < \cdots < j_q \leq n}} \omega_{i_1 \cdots i_p j_1 \cdots j_q} \omega^{i_1} \wedge \cdots \wedge \omega^{i_p} \wedge \bar{\omega}^{j_1} \wedge \cdots \wedge \bar{\omega}^{j_q}.$$

Then, every differential form on an almost complex manifold M is the sum of bihomogeneous components of type (p, q).

6.4 *Remark* Every almost complex manifold M is orientable, and we shall assume that M is oriented by the orientation which makes the local differential form $\left(\frac{i}{2}\right)^n \omega^1 \wedge \bar{\omega}^1 \wedge \cdots \wedge \omega^n \wedge \bar{\omega}^n$ positive at every $m \in M$.

We shall denote by $P_{p,q}$ the map which, to every differential form ω, assigns the component of type (p, q) of ω.

Let $\{\omega^j\}$ be a system of structure forms on a neighborhood U of $m \in M$. The differential form ω^j is of type (1, 0) and the differential $d\omega^j$ of ω^j is a differential form of degree 2 and, thus, a section of $E^2(U)$. Hence, $d\omega^j$ splits into the sum of bihomogeneous elements

$$P_{0,2}\, d\omega^j, \quad P_{1,1}\, d\omega^j, \quad P_{2,0}\, d\omega^j,$$

of type (0, 2), (1, 1), and (2, 0) respectively. It follows, as an easy computation shows, that the differential $d\omega^{p,q}$ of a differential form $\omega^{p,q}$ of type (p, q) is the sum of bihomogeneous elements of type $(p-1, q+2)$, $(p, q+1)$, $(p+1, q)$, and $(p+2, q-1)$.

We put

$$\partial\omega^j = P_{2,0} d\omega^j$$

and

$$\bar{\partial}\omega^j = P_{1,1} d\omega^j.$$

In general, we shall denote by $\partial\omega^{p,q}$ and $\bar{\partial}\omega^{p,q}$ the components of type $(p+1, q)$ and $(p, q+1)$ of $d\omega^{p,q}$ for $0 \leq p, q \leq n$.

6.5 Definition *An almost complex structure is called integrable if, for any differential form ω of type* (1, 0), $P_{0,2}\, d\omega = 0$.

6.6 Theorem *Let M be an almost complex manifold. The differential $d\omega^{p,q}$ of any differential form of type (p, q) on M is the sum of two bihomogeneous forms of type $(p+1, q)$ and $(p, q+1)$ if and only if the almost complex structure of M is integrable.*

Proof Since the condition is clearly necessary, we shall prove the sufficiency. We remark first that, if $\{\omega^j\}$ are the local structure forms, then $P_{0,2}\, d\omega^j = 0$ and, thus, $P_{2,0}\, d\bar{\omega}^j = 0$. In order to show that, if $\omega^{p,q}$ is any differential form of type (p, q), $d\omega^{p,q}$ is the sum of a form $\partial\omega^{p,q}$ of type $(p+1, q)$ and of a form $\bar{\partial}\omega^{p,q}$ of type $(p, q+1)$, it suffices to show that this is true for an element

$$\omega^{i_1} \wedge \cdots \wedge \omega^{i_p} \wedge \bar{\omega}^{j_1} \wedge \cdots \wedge \bar{\omega}^{j_q}.$$

This follows at once from the above remark, and we leave this verification to the reader. ○

It follows from the previous theorem that, if the almost complex structure of M is integrable, then $d = \partial + \bar{\partial}$ and

$$\partial^2 = 0, \quad \bar{\partial}^2 = 0, \quad \partial\bar{\partial} + \bar{\partial}\partial = 0.$$

Thus, if M is an almost complex manifold and the almost complex structure is integrable, we can operate formally on differential forms as in the case of complex manifolds.

If, now, M is a complex manifold, holomorphic functions f on M may be defined by the condition $\bar{\partial}f = 0$ where $\bar{\partial}$ is defined by the almost complex structure induced on M by its complex structure. It follows that an almost complex structure on M is induced by, at most, a unique complex structure on M.

A necessary condition, which follows from theorem 6.6, for an almost complex structure to be induced by a complex structure is that the almost complex structure be integrable. It can be proved [13] that this condition is also sufficient. Thus, on a differentiable manifold M there is no difference between complex structures and integrable almost complex structures compatible with the given differentiable structure of M. The proof of this theorem is delicate and we omit it.

The question of the existence of almost complex structures on even-dimensional orientable differentiable manifolds is a very difficult topological problem. Every complex manifold is, of course, an almost complex manifold and theorem 6.6 shows that the converse is not necessarily true.

7.1 The fundamental form

Let M be a Hermitian manifold and let

$$ds_\alpha^2 = \sum_{j,k} h_{jk}\, dw_\alpha^j\, d\bar{w}_\alpha^k$$

be the expression of the metric on a coordinate neighborhood (U_α, w_α). On the same neighborhood U_α, we define a differential form Ω_α of type (1, 1) by

$$\Omega_\alpha = \frac{i}{2} \sum_{j,k} h_{jk}\, dw_\alpha^j \wedge d\bar{w}_\alpha^k.$$

It is easy to verify that:

a) the differential forms Ω_α define a global differential form Ω on M; that is, Ω_α, Ω_β agree on $U_\alpha \cap U_\beta$,

b) the differential form Ω is real; that is, the coefficients of Ω are real when Ω is expressed in terms of a real basis.

7.2 DEFINITION *The differential form Ω is called the fundamental form of the Hermitian structure of M.*

Now let M be an n-dimensional almost complex manifold and let Ω be a real differential form of type (1, 1) defined on M. Let $\omega^1, \dots, \omega^n$ be the structure forms of M on a neighborhood of $m \in M$ and let $\omega^j = \xi^j + i\eta^j$, $j = 1, \dots, n$. Since

$$d\xi^j \wedge d\eta^k + d\xi^k \wedge d\eta^j = \frac{i}{2}(d\omega^j \wedge d\bar{\omega}^k + d\omega^k \wedge d\bar{\omega}^j),$$

it follows that, locally, the differential form Ω can always be written as

$$\Omega = \sum_{j,k} h_{jk}\, d\omega^j \wedge d\bar{\omega}^k$$

where the h_{jk} are complex valued differentiable functions and the condition that Ω be real is equivalent to

$$h_{jk} = \overline{h_{kj}}.$$

Thus, the assignment of a real differential form Ω of type (1, 1) on M determines at every $m \in M$, via the local expression of Ω, a Hermitian form on the cotangent space T_m^*. If this Hermitian form is positive definite at every $m \in M$, then it induces a Hermitian metric on M by

$$ds^2 = \sum_{j,k} h_{jk}\omega^j\bar{\omega}^k.$$

We shall say that the fundamental form Ω is *positive definite* if the induced Hermitian form is positive definite.

An almost complex manifold together with a positive definite fundamental form will also be called a Hermitian manifold, although it may not be a complex manifold.

Let M be a Hermitian manifold. We shall use the fundamental form Ω of the structure of M to define a local bihomogeneous operator

$$L: E^{p,q}(M) \to E^{p+1,q+1}(M)$$

by

$$L\omega = \Omega \wedge \omega$$

for every $\omega \in E^{p,q}(M)$.

The adjoint of the operator L is the bihomogeneous operator

$$\Lambda: E^{p,q} \to E^{p-1,q-1},$$

defined by

$$\Lambda \omega^{p,q} = (-1)^{p+q} * L * \omega^{p,q}$$

for every $\omega^{p,q} \in E^{p,q}(M)$.

7.3 *Remark* The operators L and Λ are defined on the fibers of $E^{p,q}(M)$ and, thus, they are linear with respect to the differentiable functions.

7.4 DEFINITION *A differential form $\omega^{p,q}$ is called primitive if $\Lambda\omega^{p,q} = 0$.*

We shall denote by $A^{p,q}$ the kernel of the map $\Lambda : E^{p,q} \to E^{p-1,q-1}$; that is, we put

$$A^{p,q} = \{\omega^{p,q} \in E^{p,q} \mid \Lambda\omega^{p,q} = 0\}.$$

7.5 THEOREM *Let* $\dim_{\mathbf{R}} M = 2n$. *If $p + q > n$, the map Λ: $E^{p,q}(M) \to E^{p-1,q-1}(M)$ is injective.*

Proof Because of remark 7.3, it suffices to prove the theorem for $\omega^{p,q}(m)$ and $\Omega^{p,q}(m)$ at a fixed point $m \in M$. In this case, the proof requires only a computation, which we leave to the reader. ○

We remark that the above theorem says that there are no primitive differential forms of type (p, q) if $p + q > n$.

8.1 Kähler manifolds

Let M be a complex manifold and let a Hermitian metric be defined on M. Let Ω denote the fundamental form of the Hermitian structure of M.

8.2 DEFINITION *A Hermitian metric is called a Kählerian metric if the fundamental form Ω is closed, that is, if $d\Omega = 0$.*

We remark that, since Ω is bihomogeneous, the condition $d\Omega = 0$ implies that $\partial\Omega = 0$ and $\bar{\partial}\Omega = 0$.

8.3 DEFINITION *A complex manifold M is called a Kähler manifold if M can be given a Kählerian metric.*

Not every complex manifold can be given a Kählerian metric. The next theorem shows a necessary condition for a compact complex manifold to be a Kähler manifold.

8.4 THEOREM *Let M be a compact Kähler manifold,* $\dim_{\mathbf{C}} M = n$. *Then,*

$$H^{2p}(M, \mathbf{R}) \neq 0$$

for $p = 0, 1, \ldots, n$.

Proof If Ω is the fundamental form of a Kählerian metric on M, we shall denote by $\Omega^{(p)} = \Omega \wedge \cdots \wedge \Omega$ the p-fold exterior product of Ω by itself.

From $d\Omega = 0$ it follows $d\Omega^{(p)} = 0$; hence, the differential form $\Omega^{(p)}$ represents, via de Rahm's theorem, an element of $H^{2p}(M, \mathbf{R})$. To prove the theorem, we must show that $\Omega^{(p)}$ is not the differential of a differential form of degree $2p - 1$.

Assume there is a differential form η^{2p-1} such that

$$\Omega^{(p)} = d\eta^{2p-1}.$$

Then we would have

$$\Omega^{(n)} = d(\Omega^{(n-p)} \wedge \eta^{2p-1})$$

and, thus, by Stokes' theorem, ○

$$\int_M \Omega^{(n)} = \int_M d(\Omega^{(n-p)} \wedge \eta^{2p-1}) = 0.$$

But a straightforward computation shows that

$$\Omega^{(n)} = 2^n \det(h_{jk})\, dV,$$

where dV is the positive volume element of M. Thus,

$$\int_M \Omega^{(n)} > 0.$$

This contradiction proves the theorem.

8.5 *Remark* There are examples of complex manifolds which do not satisfy the thesis of theorem 8.4. One example is the Hopf manifold $\mathscr{H}^2$. $\mathscr{H}^2$ is a complex manifold of complex dimension 2 which is topologically equivalent to the product $S^1 \times S^3$. It can be proved that $H^2(\mathscr{H}^2, \mathbf{R}) = 0$; thus, the Hopf manifold $\mathscr{H}^2$ is not a Kähler manifold.

8.6 Theorem *Let M be a Kähler manifold and let $\phi : N \to M$ be a regular holomorphic map of a complex manifold N into M. Then N is a Kähler manifold.*

Proof Let Ω be the fundamental form of the Kählerian structure of M. The metric on M induces a Hermitian metric on N. The fundamental form of the induced metric is the inverse image $\phi^*\Omega$ of Ω. Since

$$d(\phi^*\Omega) = \phi^*(d\Omega) = 0,$$

the induced metric on N is Kählerian. ○

8.7 THEOREM *Let M be a Kähler manifold and let Ω be the fundamental form of the structure of M. To every point $m \in M$ there exist a neighborhood $U \subset M$ of m and a real valued differentiable function f on U such that*

$$\Omega = i\partial\bar{\partial} f.$$

Proof The differential form Ω is real and $d\Omega = 0$. By Poincaré's lemma 8.3.2, there exists locally a real differential form ω of degree 1 such that $\Omega = d\omega$. This differential form ω is the sum of two bihomogeneous differential forms of type (1, 0) and (0, 1) respectively. Because ω is real, these two components must be conjugate; thus, we can write

$$\omega = \eta + \bar{\eta}.$$

Then we have

$$\Omega = d\omega = d(\eta + \bar{\eta}) = \partial\eta + \bar{\partial}\eta + \partial\bar{\eta} + \bar{\partial}\bar{\eta}.$$

Since Ω is of type (1, 1), we must have $\partial\eta = \bar{\partial}\bar{\eta} = 0$, and, thus,

$$\Omega = \bar{\partial}\eta + \partial\bar{\eta}.$$

Now the condition $\bar{\partial}\bar{\eta} = 0$ implies locally, by Dolbeault's lemma 8.5.4, the existence of a complex valued differentiable function ϕ such that $\bar{\eta} = \bar{\partial}\phi$. By conjugation, we also have $\eta = \partial\bar{\phi}$.

Then we have

$$\Omega = \bar{\partial}\partial\bar{\phi} + \partial\bar{\partial}\phi = \partial\bar{\partial}(\phi - \bar{\phi}),$$

and, if we put $f = \frac{1}{2}(\phi - \bar{\phi})$, we have

$$\Omega = i\partial\bar{\partial} f. \bigcirc$$

Now let M be a complex manifold and assume that Ω is a real differential form of type (1, 1) on M with the property that, on a neighborhood of every point of M, there is a real valued differentiable function f such that $\Omega = i\partial\bar{\partial}f$. Because

$$d\Omega = i\partial\partial\bar{\partial}f + i\bar{\partial}\partial\bar{\partial}f = -i\partial\bar{\partial}\bar{\partial}f = 0,$$

such a differential form is closed. We may ask when a differential form Ω defined on M and locally of the form $i\partial\bar{\partial}f$ is the fundamental form of a Kählerian structure on M. We have only to check whether the quadratic form associated with Ω is positive definite. The next theorem provides some information to this end.

8.8 THEOREM *Let M be an n-dimensional complex manifold and let $f_1, \ldots, f_r$ be $r > n$ holomorphic functions on an open set $U \subset M$ such that:*

i) *the functions f_j do not vanish together at any point of U,*

ii) *at every point $m \in U$, where $f_{j_0}(m) \neq 0$, there are n linearly independent differential forms among the*

$$d(f_1/f_{j_0}), \ldots, d(f_r/f_{j_0}).$$

Then the real valued differentiable function

$$f = \log\left(\sum_{j=1}^{r} f_j \bar{f}_j\right)$$

defines a differential form

$$\Omega = i\partial\bar{\partial}f$$

which is positive definite on U.

Proof Since the f_j's are holomorphic, we have

$$df_j = \partial f_j, \quad \bar{\partial} f_j = 0,$$

and

$$d\bar{f}_j = \bar{\partial}\bar{f}_j, \quad \partial\bar{f}_j = 0.$$

It follows that

$$\Omega = i\left(\sum_j f_j \bar{f}_j\right)^{-2}\left[\left(\sum_j f_j\bar{f}_j\right) \wedge \left(\sum_j df_j\, d\bar{f}_j\right) - \left(\sum_j \bar{f}_j\, df_j\right) \wedge \left(\sum_j f_j\, d\bar{f}_j\right)\right]$$

$$= \frac{i}{2}\left(\sum_j f_j \bar{f}_j\right)^{-2} \sum_{j,k} (f_j\, df_k - f_k\, df_j) \wedge (\bar{f}_j\, d\bar{f}_k - \bar{f}_k\, d\bar{f}_j).$$

Let $m \in U$, let $t \in T_m$, and let

$$\omega_{jk}(t) = f_j(m)\,\langle t, df_k\rangle - f_k(m)\,\langle t, df_j\rangle.$$

Then, the Hermitian form

$$H = \left(\sum_{j=1}^{r} f_j \bar{f}_j\right)^{-2} \sum_{j,k} \omega_{jk}\bar{\omega}_{jk}$$

is certainly positive definite, unless the ω_{jk} are all equal to zero. We leave it to the reader to verify that, if $f_{j_0}(m) \neq 0$, this can happen if and only if there are no n linearly independent differential forms among the

$$d(f_1/f_{j_0}), \ldots, d(f_r/f_{j_0}). \quad \bigcirc$$

9.1 Examples

9.2 **C**^{n} *is a Kähler manifold* The Euclidean metric in **C**^{n},

$$ds^2 = \sum_{j=1}^{n} dz^j \, d\bar{z}^j,$$

is Hermitian, and the fundamental form has the expression

$$\Omega = i \sum_{j=1}^{n} dz^j \wedge d\bar{z}^j.$$

Thus, $d\Omega = 0$. We remark that, in **C**^{n}, there exists a global real valued differentiable function $f = \sum_{j=1}^{n} z_j \bar{z}_j$ such that $\Omega = i\partial\bar{\partial} f$.

9.3 *Riemann surfaces are Kähler manifolds* Actually, in this case, every Hermitian metric is Kählerian. In fact, if Ω is the fundamental form of any Hermitian structure of a Riemann surface M, then the degree of $d\Omega$ is $3 > 2 = \dim_{\mathbf{R}} M$; thus, $d\Omega = 0$.

9.4 *Stein manifolds are Kähler manifolds* This follows from theorem 8.6 with $M = \mathbf{C}^n$.

9.5 **P**^{n}(**C**) *is a Kähler manifold* Let $z_0, \ldots, z_n$ be coordinates of **C**$^{n+1}$. For every j, $0 \le j \le n$, consider the open set $U_j = \{z_j \neq 0\}$. The collection $\{U_j\}$ is an open covering of **P**^{n}(**C**).

On the open set U_j, we consider the $n + 1$ holomorphic functions

$$z_0/z_j, \ldots, z_n/z_j.$$

Among these functions there is the constant function $z_j/z_j = 1$; thus, they do not vanish together on U_j. The remaining functions are a system of local coordinates on U_j. Therefore, their differentials $d(z_0/z_j), \ldots, d(z_n/z_j)$ are linearly independent at every point of U_j. Thus, by theorem 8.8, the function

$$f_j = \log \sum_{k=0}^{n} |z_k/z_j|^2$$

defines a differential form

$$\Omega_j = i\partial\bar{\partial} \log \sum_{k=0}^{n} |z_k/z_j|^2$$

which is positive definite on U_j.

Now, an easy computation shows that on the intersection $U_j \cap U_l$ we have

$$\Omega_j - \Omega_l = i\partial\bar{\partial} \log |z_l/z_j|^2 = 0.$$

This proves the existence of a unique, closed, positive definite differential form Ω on $\mathbf{P}^n(\mathbf{C})$ such that, for every $j = 0, \ldots, n$, $\Omega | U_j = \Omega_j$. Thus, Ω defines a Kählerian structure on $\mathbf{P}^n(\mathbf{C})$.

9.6 *Every complex projective algebraic variety is Kähler* This follows from 9.5 and theorem 8.6.

9.7 *Complex tori are Kähler manifolds* A complex torus is a quotient space of $\mathbf{C}^n$ by a translation group. The Euclidean Hermitian metric on $\mathbf{C}^n$ is invariant under translations and, therefore, induces a Hermitian structure on the torus. Thus, every point of a complex torus has a neighborhood where the induced Hermitian metric coincides with that of $\mathbf{C}^n$.

10.1 Holomorphic and harmonic forms

We recall that, if S, T are any two operators, the bracket $[S, T]$ is defined by $[S, T] = ST - TS$. Thus, if the operators commute, we have $[S, T] = 0$.

10.2 LEMMA *On a Kähler manifold we have:*

a) $[L, d] = [L, \partial] = [L, \bar{\partial}] = 0,$
b) $[\Lambda, \delta] = [\Lambda, \bar{\theta}] = [\Lambda, \theta] = 0.$

Moreover,

c) $[L, \theta] = -i\partial, \; [L, \bar{\theta}] = i\bar{\partial},$

d) $[\Lambda, \bar{\partial}] = -i\theta, \quad [\Lambda, \partial] = i\bar{\theta}.$

Proof We prove the first of relations (a). If $d\Omega = 0$, then, for every differential form ω, we have

$$dL\omega = d(\Omega \wedge \omega) = \Omega \wedge d\omega = L\, d\omega.$$

We leave the other similar proofs to the reader. ○

10.3 LEMMA *On a Kähler manifold*

$$(\partial\theta + \theta\partial) = (\bar{\partial}\bar{\theta} + \bar{\theta}\bar{\partial}) = 0.$$

Proof We shall use the last two relations of lemma 10.2. We have

$$\partial\theta = -i\partial(\Lambda\partial - \partial\Lambda) = -i\partial\Lambda\partial,$$

$$\theta\partial = -i(\Lambda\partial - \partial\Lambda)\partial = i\partial\Lambda\partial.$$

Thus, $\partial\theta + \theta\partial = 0$.

Similarly,

$$\bar{\partial}\bar{\theta} = i\bar{\partial}(\Lambda\bar{\partial} - \bar{\partial}\Lambda) = i\bar{\partial}\Lambda\bar{\partial},$$

$$\bar{\theta}\bar{\partial} = i(\Lambda\bar{\partial} - \bar{\partial}\Lambda)\bar{\partial} = -i\bar{\partial}\Lambda\bar{\partial}.$$

Thus, $\bar{\partial}\bar{\theta} + \bar{\theta}\bar{\partial} = 0$. ○

10.4 LEMMA *On a Kähler manifold* $\square' = \square''$.

Proof By definition, $\square' = \partial\bar{\theta} + \bar{\theta}\partial$ and $\square'' = \theta\bar{\partial} + \theta\bar{\partial}$. Now, by lemma 10.2,

$$\partial\bar{\theta} + \bar{\theta}\partial = i[\partial(\Lambda\bar{\partial} - \bar{\partial}\Lambda) + (\Lambda\bar{\partial} - \bar{\partial}\Lambda)\partial]$$

$$= i(\partial\Lambda\bar{\partial} - \partial\bar{\partial}\Lambda + \Lambda\bar{\partial}\partial - \bar{\partial}\Lambda\partial),$$

and

$$\bar{\partial}\theta + \theta\bar{\partial} = i[-\bar{\partial}(\Lambda\partial - \partial\Lambda) - (\Lambda\partial - \partial\Lambda)\bar{\partial}]$$

$$= i(-\bar{\partial}\Lambda\partial + \bar{\partial}\partial\Lambda - \Lambda\partial\bar{\partial} + \partial\Lambda\bar{\partial});$$

thus, $\partial\bar{\theta} + \bar{\theta}\partial = \bar{\partial}\theta + \theta\bar{\partial}$. ○

10.5 THEOREM *On a Kähler manifold, the real Laplace operator* Δ *is bihomogeneous of type* (0, 0). *Moreover,*

$$\Delta = 2\square' = 2\square''.$$

Proof By lemmas 10.3 and 10.4, we have

$$\Delta = d\delta + \delta d = (\partial + \bar{\partial})(\theta + \bar{\theta}) + (\theta + \bar{\theta})(\partial + \bar{\partial})$$

$$= (\partial\bar{\theta} + \bar{\theta}\partial) + (\bar{\partial}\theta + \theta\bar{\partial}) = \square' + \square''. \;\bigcirc$$

It follows from lemma 10.4 that, on a Kähler manifold, we can put $\square = \square' = \square''$. Moreover, by theorem 10.5, $\square$ is a real operator. Also, by the same theorem, on a Kähler manifold there is no difference between complex valued Δ-harmonic and $\square$-harmonic differential forms.

10.6 THEOREM *Let* M *be a compact Kähler manifold and let* η *be a differential form of type* $(p, 0)$ *on* M. *The following conditions are equivalent:*

i) η *is d-closed,*
ii) η *is holomorphic,*
iii) η *is harmonic.*

Proof (i) implies (ii) on every complex manifold. On Kähler manifolds, by theorem 10.5, $\Delta = 2\Box = 2(\bar{\partial}\theta + \theta\bar{\partial})$. Because η is of type $(p, 0)$, we have $\theta\eta = 0$ and, thus, $\Delta\eta = 2\theta\,\bar{\partial}\eta$, which shows that (ii) implies (iii). On every compact manifold, by theorem 4.3, condition (iii) implies (i). ○

11.1 Problems

1 (1.1) Prove properties (i)–(vi) of the operator $*$.

2 (4.1) Prove that $[*, \Delta] = [d, \Delta] = [\delta, \Delta] = 0$.

3 (5.1) Show that every complex manifold can be given a Hermitian metric.

4 (5.1) Show that the assignment of a Hermitian metric on a complex manifold M is equivalent to a reduction of the structure group of the holomorphic tangent bundle $H(M)$, as a complex-differentiable bundle, to the unitary group.

5 (7.1) Prove that the fundamental form Ω of a Hermitian structure is real.

6 (7.1) Complete the proof of theorem 7.5.

7 (7.1) State and prove a theorem analogous to theorem 7.5 for the operator L.

8 (8.1) Fill in the missing details in the proof of theorem 8.8.

9 (10.1) Prove all relations in lemma 10.2.

Bibliography

1. Auslander, L., and MacKenzie, R.E., *Introduction to Differentiable Manifolds*, McGraw-Hill, New York (1963).
2. Bishop, R.L., and Crittenden, R.J., *Geometry of Manifolds*, Academic Press, New York (1964).
3. Bredon, G.E., *Sheaf Theory*, McGraw-Hill, New York (1967).
4. Chern, S.S., *Complex Manifolds*, University of Chicago lecture notes (1956).
5. Chern, S.S., *Differentiable Manifolds*, University of Chicago lecture notes (1959).
6. Chevalley, C., *Theory of Lie Groups*, Princeton University Press, Princeton, N.J. (1946).
7. Dolbeault, P., "Formes différentielles et cohomologie sur une variété analytique complexe", *Ann. Math.* (1956), (1957).
8. Godement, R., *Topologie Algébrique et Théorie des Faisceaux*, Hermann, Paris (1958).
9. Gunning, R.C., and Rossi, H., *Analytic Functions of Several Complex Variables*, Prentice-Hall, Englewood Cliffs, N.J. (1965).
10. Hausemoller, D., *Fibre Bundles*, McGraw-Hill, New York (1966).
11. Hopf, H., *Zur Topologie der komplexen Mannigfaltigkeiten*, Interscience, New York (1948).
12. Kervaire, M., "A manifold which does not admit any differentiable structure", *Comm. Math. Helv.*, 35 (1961).
13. Newlander, A., and Nirenberg, L., "Complex analytic coordinates in almost complex manifolds", *Ann. Math.*, 65 (1967).
14. de Rahm, G., *Variétés Différentiables*, Hermann, Paris (1955).
15. Seminario E.E. Levi, *Lecture Notes*, University of Pisa (1962).
16. Steenrod, N., *The Topology of Fibre Bundles*, Princeton University Press, Princeton, N.J. (1951).
17. Weil, A., *Variétés Kählériennes*, Hermann, Paris (1958).

Subject index